BEYOND LIGHT BULBS

LIGHTING THE WAY TO SMARTER ENERGY MANAGEMENT

Susan Meredith

EMERALD
BOOK CO.

Published by Emerald Book Company
4425 S. Mo Pac Expy., Suite 600
Austin, TX 78735

Distributed by Emerald Book Company

For ordering information or special discounts for bulk purchases, please
contact Emerald Book Company at 4425 S. Mo Pac Expy., Suite 600,
Austin, TX 78735, (512) 891-6100.

Design and composition by Greenleaf Book Group LLC
Cover design by Ian Shimkoviak and Greenleaf Book Group LLC

Publisher's Cataloging-In-Publication Data
(Prepared by The Donohue Group, Inc.)

Meredith, Susan (Susan Elizabeth), 1958-
    Beyond light bulbs : lighting the way to smarter energy management / Susan Meredith. -- 1st ed.

    p. : charts ; cm.

    ISBN: 978-1-934572-07-8

1. Energy conservation. 2. Energy consumption. I. Title.

TJ163.3 .M47 2009
333.7   2008932861

Printed in the United States of America on acid-free paper

08 09 10 11 12 13 14   10 9 8 7 6 5 4 3 2 1

First Edition

*For the Earth and
all that live here.*

CONTENTS

INTRODUCTION

Almost two years ago, I set out to determine how the average citizen can learn enough about global warming and the energy crisis to make a difference. This book is the result of my quest.

I thought I was fairly knowledgeable about energy and the environment. When I was at the University of Illinois–Champaign Urbana in 1980, getting my engineering degree, I completed a project on electric cars. At the time, the post office in Schaumburg used them for mail-delivery trucks.

In subsequent years, I became acquainted with an organization called Global Energy Network Institute (GENI), which promoted the beliefs of Buckminster Fuller, the noted inventor, author, architect, and futurist. The geodesic dome, which Fuller invented, became a model for fundamental structures in nanotechnology called fullerenes and buckyballs. He also invented a game designed to promote peace, called the World Game, which posed this question: How can we make the world work for 100 percent of humanity in the shortest possible time through spontaneous cooperation without ecological damage or the disadvantage of anyone?

Fuller had arrived at the conclusion that a global energy grid was the key to solving world problems, including hunger, overpopulation, and even wars. At various times over the past few years, I have spoken on GENI's behalf to groups as large as three hundred and assisted the director, Peter Meisen, at the World Energy Congress in Houston in 1998, which President George H. W. Bush attended.

But when I was earning my master's in business administration, through the Texas Evening MBA program at the University of Texas–Austin (I graduated in 2006), I realized how much I had yet to learn about global energy.

In the final year of the program, they hold a capstone project in which teams usually solve a business problem and present the results to a "board of directors." Our project was on how to move to the hydrogen economy, based on reference documentation we were provided.[1] With eighty of some of the brightest MBA students in the country competing for a prize of one thousand dollars for each team member, that was a lot of brainpower focused on solutions to global energy. It occurred to me that this information shouldn't be wasted. I gathered up research from the other students and, with the aid of the exceptional Milap Majmundar, a member of the winning team, furthered the research and created the core of this book.

In addition, I was fortunate to come across Mark Kapner at Austin Energy, in Austin, Texas, the nation's ninth largest community-owned electric utility. Austin Energy created the top-performing renewable energy program in the nation and owns the nation's first and largest green building program. Mark is one of those wonderful individuals who has carried the torch for renewable energy for years. He was extremely generous with his knowledge and was an amazing resource.

I sent the resulting forty-six–page report to the Department of Energy and the Council on Environmental Quality in April 2006. I also presented our findings—which extended well beyond hydrogen as an energy solution—to the Renewable Energy Roundup and Green Living Fair, an annual showcase of speakers and exhibitors, in Fredericksburg, Texas, in September 2006.

My interest was stirred. I began to attend the CleanTX Forums hosted by the Clean Energy Incubator in Austin, and I looked for other opportunities to learn and share information, both formally and informally.

I attended the last two weeks of the Texas legislative session in May 2007 to learn about policies and standards. I was extremely fortunate to make the acquaintance of a retired schoolteacher named Dwight Harris, who had taught government. I met him in the gallery of the House of Representatives. Though I was a complete stranger, Dwight was very generous with his knowledge. He helped me unravel the mysteries of a bill's passage through the legislative process. Soon, I became an unpaid lobbyist of sorts, talking to as many people as possible about the importance of laws that support alternative energy, and even testifying during a meeting of the legislature's Business and Commerce Committee about a renewable-energy bill.

During this time, I was having difficulty reconciling my day job with my passion for energy. HumanExcel, the corporate education company I founded in 1989, focuses on people, process, and metrics. My own personal work focused on improving business operations using process analysis and statistical tools. At some point, it dawned on me that the only disconnect between the two areas was in my mind. Truly, what I had been teaching for years to my corporate clients was organizational energy management and how to minimize waste and use resources efficiently on an organizational level. I realized, also, that the team aspects of our company really dealt with human energy management on a personal level. This discovery has been invaluable in developing a framework for addressing the global energy crisis.

This book has grown to encompass all aspects of global energy and more. There are graphs and numbers for those who like graphs and numbers. If you don't like them, you don't need to look at them. You will still understand the concepts.

I don't consider myself a foremost authority on global energy or renewable technologies or electricity. There are so many brilliant people who are, and I appreciate how willing many of them have been in sharing their knowledge with me. But my lack of specific expertise is what makes me qualified, I think, to write this book. I am a generalist on energy management. I have no attachment to any particular product, solution, or organization. I won't get heavily into techno-speak. I have tried to write this book for the person who is not an expert yet wants to understand, in simple terms, what he or she can do to address this problem.

I firmly believe that together we can solve the problems on this planet. My hope is that this book helps us get there.

THE ENERGY MANAGEMENT OPPORTUNITY

"WHEN ARE THEY GOING to make a movie that has a positive future?" my husband asked after watching yet another futuristic film showing destroyed buildings, chaos, violence, confusion, and a pervasive mood of despair. How depressing! Yet, when I thought about it, I realized that the prevailing mood in the real world is often depressing: We're doomed … But just in case, change your light bulbs.

Yeah, right.

I had this nagging thought that the "Ignore it and maybe it will go away" approach wasn't wise. From business schools to Oprah shows, the message is the same: Create a vision of the future. Focus on what you want to generate, not what you want to avoid. Problems create opportunities.

Well, then, maybe it was time to start looking for opportunities to do something …

ENERGY MANAGEMENT

Everything is made up of energy. We are made of energy. The Earth is made of energy. Even our thoughts and feelings have energy.

Our primary focus here will be on global energy. Global energy generally means the kinds of energy that heat and cool our homes and workplaces and fuel our vehicles. Yet the principles of energy management apply on all levels: global, personal, and organizational.

Energy management means managing energy so that it is channeled in the right direction, along the most efficient paths, and distributed at a rate that supplies the right amount of energy at the right time to meet the demand. Energy managed so that it isn't wasted. Energy regulated so that it doesn't harm us. Energy that doesn't pollute our environment. Energy that produces results.

On a global scale, our largest use of energy is in the form of electricity. In addition, we use energy to transport food, building supplies, other goods, and people.

Without appropriate energy, our commercial, educational, health, finance, transportation, and defense systems cannot operate. On a more personal level, without energy we wouldn't have refrigerators for our food, lights for our homes, clean water on demand, or TVs and other electronics for our entertainment. Appliances such as washers and dryers, toasters and microwaves, would no longer provide the convenience we are accustomed to.

Energy management applies to more than just those physical forms of energy that fuel our homes and propel our vehicles. It relates to human energy—the energy each of us intakes and outputs to function in our daily lives on a physical, a mental, and an emotional level. On a personal and organizational level, energy management means channeling our own energy in the right direction, along the most efficient paths, and distributing it at a rate that supplies the right amount of energy at the right time to meet the demand. Managing our own energy so that it isn't wasted. Regulating it so that it doesn't harm us or anyone around us. Ensuring that the energy we put out doesn't pollute our social environments. Using energy to produce results.

We need to use our personal and organizational energy to solve the issues of global energy. Energy management on all levels is the key to creating health for our planet and ourselves.

WORLDWIDE OPPORTUNITY

On a global level, we have some problems. It's uncomfortable to see those videos of starving children in Africa. It's depressing and sometimes frightening to hear about war, hurricanes, earthquakes, and disease. It's much easier to turn the channel or turn the page, to turn off and tune out. And time goes on.

But the problems are still there. We could argue about the urgency of reducing oil dependence. We could debate whether global warming is causing more hurricanes and typhoons. We could figure that the economy and our global relationships are doing just fine. But what if we could improve them? What if we could

- reduce dependence on foreign oil?
- minimize global warming?
- stimulate the economy?
- promote global relationships?
- reduce world hunger?

Rather than address energy as a problem, it is addressed here as an opportunity. A solution is mapped out. It is not the only solution. It is *a* solution. It is not even a solution that I am wholly advocating. It is primarily used to show the components and how they could fit together. The proposals are not unique. What is unique, perhaps, is that this book shows a pathway to get there.

Solving the energy problems does not have to mean deprivation. There is neither a scarcity nor an overabundance of energy. As a matter of fact, the amount of energy in our universe doesn't change. One of the laws of physics states that energy is neither created nor destroyed; it just changes form. The challenge is in getting the energy in the form we need at the appropriate time and place. We now have the opportunity to assess energy management on a global scale and to create a solution that works for everyone.

DEPENDENCE ON OIL

I'D LIKE TO PRETEND that my motivation for focusing on energy solutions is entirely altruistic and high-minded, but that's just not the case. It was sparked when I went back to school to get my master's in business administration. Twenty-five years since I'd been a student, the oldest of the eighty students in the program by quite a few years—it was exciting, intimidating, and invigorating. Exhausting, too! I was also running a business and trying to find time for my then five-year-old son and my husband.

I went into the capstone competition about moving to a hydrogen economy with a bias toward renewable energy. Another member of our team was a tried-and-true gas-and-oilman with a father in the industry. Both of us, as well as the other four members of the team, learned more than we could have imagined.

One full week in August 2005 was dedicated, day and night, to coming up with a solution. As I quickly learned, there were many basic concepts that I needed to grasp before I could even get started. Like, What's the difference among oil, gas, and natural gas? and How do you convert gallons or barrels of oil to BTUs and kilowatts?

By definition, our project required a hydrogen solution. Based on subsequent research, I've proposed a different transition with a balanced energy solution.

SOURCES AND USES OF OIL

It's no secret that the United States is dependent on oil. In addition to domestic sources, our largest suppliers in 2007 were Canada, Mexico, Venezuela, and Saudi Arabia (see figure 2-1). Sixty percent of our oil is imported.[1]

In recent years, oil prices have risen precipitously. They went from forty dollars per barrel in 2004 to one hundred dollars in 2007—more than double in only three years.

The price of oil affects much more than the price of gas at the pump. Oil and the energy it provides are used for transporting goods, manufacturing products, lighting office buildings, and heating our homes. Their by-products are also used to make plastic bottles, bags, and toys. They

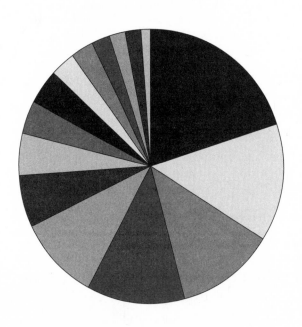

20% Canada
14% Mexico
12% Venezuela
12% Saudi Arabia
10% Nigeria
 6% Algeria
 5% Angola
 4% Iraq
 4% Russia
 3% Virgin Islands
 3% United Kingdom
 2% Brazil
 2% Ecuador
 2% Kuwait
 1% Colombia

FIGURE 2-1: Total Imports of Petroleum, 2007

are used for fertilizer, pesticides, and the asphalt on our roads. They are even used in pharmaceuticals (see figure 2-2). What this means to the consumer is that when the price of oil rises, the price of virtually everything rises, and we feel the pinch in our pocketbooks.

Another concern is a phenomenon called peak oil. Peak oil is a term used to describe the fact that oil will eventually reach its peak production capability, or supply, and we will then be unable to keep up with the increasing demand. Our use of oil has continued to rise throughout the years, and production has continued to rise to accommodate it. However, since oil is a nonrenewable substance, there is a finite amount of it underground. And we aren't the only ones whose need for it is increasing. With China and India in high-growth mode, their need for oil is increasing even faster than ours.

While the demand for oil is still going up, there are concerns that peak production capability is not far off. Once that peak is reached, the amount that can be produced not only will stop rising but also will start going down. This peak oil phenomenon contributes to the concern about our dependence on oil. There is great disagreement about whether peak

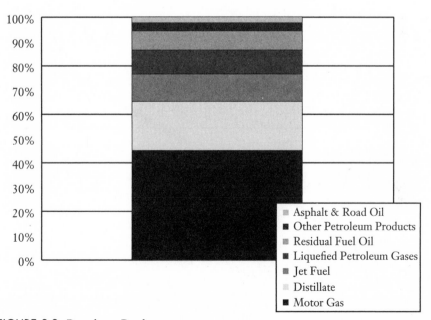

FIGURE 2-2: Petroleum Products

capacity for oil production will be reached soon. Nevertheless, the fact is that it will happen eventually because, as I said, oil is nonrenewable.

The United States currently has 22 billion barrels of oil reserves. The Middle Eastern countries referred to previously have 636 billion barrels. We aren't able to see their reserves, though, and must trust that they truly exist. There is extensive research indicating that this may not be the case.[2] In addition, global political instability has led to an uncertain supply chain and a very uneasy situation. This is a crisis that people are rightly concerned about.

So let's look at what we can do about it. As shown in figure 2-3, the largest opportunity to reduce our dependence on oil is with motor gasoline.[3]

Motor gasoline is that liquid we put into our cars, trucks, and SUVs. It is not the same as natural gas, which is used in gas stoves and for heating our homes. In Europe they call motor gas "petrol." For the purposes of this document, I will distinguish between the two substances by using the terms *petrol gas* and *natural gas*.

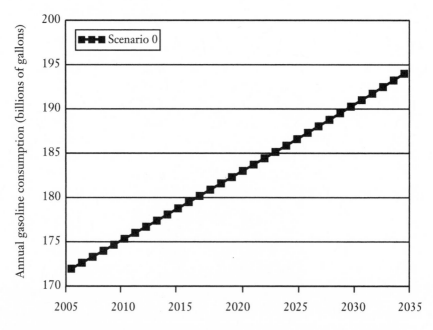

FIGURE 2-3: Gasoline Required to Fuel All Automobiles in the United States

TRANSPORTATION

Petrol Gas Consumption

We'll start looking for opportunities to break our dependence on oil with transportation, since about half of our oil is used for petrol gas.

Figure 2-3 shows projected gasoline consumption with current transportation technologies and regular petrol gas as fuel.[4] You don't even have to look at the numbers to tell that demand is on the rise.

This increase is partly due to population growth and an increase in the number of cars per household. Consumption in the United States may level off somewhat. However, there has been a tremendous increase in transportation demand in China, India, and other developing countries—China has almost doubled its consumption in the past ten years—which naturally affects gasoline consumption. Therefore, for our purposes, this remains a realistic baseline, a starting point for comparison.

Now let's look at other possible scenarios and see how they represent a reduction in petrol gas consumption and, therefore, our oil consumption and emissions.

Alternative Vehicles

GAS/ELECTRIC   By now, most people know that there are vehicles that can be powered by both gas and electricity. The Toyota Prius is currently the most visible, but there are also gas/electric versions of other automakers' vehicles. These cars are powered by an electric motor and a rechargeable battery in addition to a gas motor. Gas consumption is significantly less than with an all-gas-powered vehicle. That's great, but with the combination of more vehicles on the roads, here and overseas, and supplies peaking, prices and emissions will continue to rise. More must be done.

PLUG-INS   What isn't as well known is that cars that run exclusively on electricity, or plug-ins, are also technically feasible. They can be plugged into a standard 110-volt outlet at night, and the savings on fuel costs are significant. At a very conservative (and possibly unlikely to be seen again!) price of $2.50 per gallon and a fuel efficiency of twenty-five miles per gallon, a traditional petrol gas car costs 10 cents per mile to operate. The current residential retail cost for a kilowatt hour (kWh) of electricity is 10 cents. Current electric cars can travel four miles per kWh, so it costs 2.5 cents per kWh, or one-fourth the cost of operating a gasoline-powered car.[5] In other words, electricity would cost an equivalent amount of 63 cents per gallon based on these assumptions. As petrol gas prices continue to rise, the savings will continue to increase.

The problem with fully electric vehicles is that they have a limited driving range—the distance they can go before needing to be recharged. Neighborhood electric vehicles, or NEVs, can run about thirty miles on electricity. While many can accelerate to higher speeds, laws prohibit speeds above 30 mph. They are currently placed in the same category as golf carts, so most don't add all of the safety features required for higher-speed vehicles. There are many delivery or maintenance vehicles with predictable routes for which these limitations would not be a problem.

The average miles per day for ordinary drivers is thirty-four, with half of their trips totaling less than twenty-five miles; thus, on most days, electricity is the only "fuel" needed. Therefore, it is possible for many two-car families to use an NEV for their local transportation. Start tracking how many miles you drive in a day.

Highway-ready electric vehicles, or EVs, meet the safety regulations for driving at higher speeds. That means they have air bags and everything else that has become required over the years. Some of these EVs are coming out with projections of two hundred miles per recharge, which eliminates fears of running out of charge while driving around town. As we will see, these all-electric vehicles become more and more important as petrol gas prices rise.

Recharging can currently take up to six hours. However, combining plug-in technology with gas or other fuel types solves this problem. For instance, for anyone who needs to have the flexibility to travel farther than a battery's capacity permits, a petrol gas/electric hybrid allows the option of switching to petrol gas when necessary.

Other possible renewable technologies include biodiesel, hydraulic, solar, and hydrogen-based cars. All of these plug-in hybrid electric vehicles, or PHEVs, would reduce the need for petrol gas. For our purposes here, Component #1 of our solution is all kinds of plug-in vehicles, including fully electric vehicles, gas/electric plug-in hybrids, and other plug-in hybrids. To simplify the calculations, and also to provide the most conservative estimate of petrol gas reduction, all vehicles are projected as petrol gas/electric plug-in hybrids.

Figure 2-4 shows the mix of vehicle types over time.[6] The number of petrol gas–only vehicles goes down. Hybrids increase slightly, but will convert to plug-ins when gas/electric plug-ins kick in. The top line is the total number of automobiles.

So why aren't the car companies making these today? The documentary *Who Killed the Electric Car?* which premiered at the Sundance Film Festival in 2006 and was released by Sony Pictures, showed that electric cars have been technically viable for many years. The movie tells the story of the EV1, a General Motors battery electric car released in California in the 1990s. However, there is work to be done to make them commercially viable. If you save on your fuel, but the price of the car

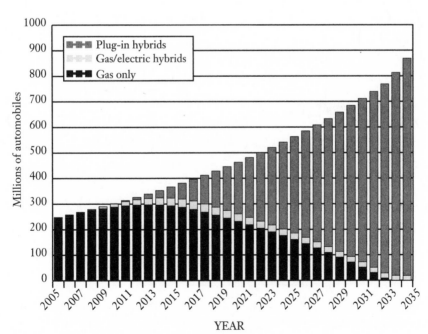

FIGURE 2-4: Automobile Types in the United States

itself is four times more than the price of existing petrol gas–fueled vehicles, most people won't be able to buy them or simply won't want to.

From a cost-analysis standpoint, it is illogical for existing car companies to initiate the move to electric cars without outside pressure. Would you take a lower-paying job if your existing job were fine and paid more? Probably not—unless the pain of staying the same exceeded the pain of changing. The current pressure to make changes comes from government regulations, from market demand, and from competitors that create an attractive alternative, as well as each car company's own desire to be part of the solution. Understandably, the car companies are responding.

Figure 2-5 compares petrol gas consumption in the United States, if we continue to use petrol-gas-burning vehicles, to U.S. consumption under alternate scenarios.

While petrol gas/electric hybrids like the Prius take us a long way toward cutting oil consumption for our vehicles, and plug-in hybrids take us even further, we can't expect all drivers to convert immediately. Even assuming all new cars are required to be plug-ins by 2015, there

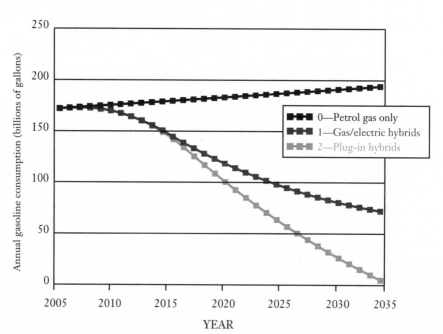

FIGURE 2-5: Gasoline Required to Fuel All Automobiles in the United States

would still be petrol gas vehicles on the road. What else can be done to reduce the demand for oil and do it more quickly?

Alternative Fuels

Component #2 of our solution is to use alternative forms of motor fuel. Consumers can convert to alternative fuels without the high cost of a new vehicle, so this is possibly the quickest way to move beyond our dependence on oil. Another significant aspect of this solution is that it can use the existing distribution infrastructure—the pumps at your local gas station.

The two most prevalent forms of alternative fuels are ethanol mixtures and biodiesel, with many more being explored. The common characteristic of alternative fuels is that they come from crops and vegetation and other sources aside from fossil fuels.

Ethanol is produced from corn, wheat, and barley. When gasoline is replaced by corn ethanol, life-cycle greenhouse gas emissions are reduced by 21.8 percent. These emissions include CO_2 as well as methane and nitrous oxide, all of which we will explore in the "Emissions" section in chapter 3.[7]

There are concerns that ethanol reduces the food supply and animal feed that come from these crops. That would mean it drives up the cost of our foods. If every acre of corn grown in the United States were used for ethanol, it would replace only 12 percent of the petrol gas we use. Plus, when crops are used for fuel, they aren't being used for food. Obviously, this is not an ultimate solution to our dependence on petrol gas but merely a temporary alternative until other solutions can ramp up.

Other types of ethanol are being developed, such as cellulosic ethanol, that are produced from the waste products of crops instead of the edible parts. Crops may also be developed that can grow on now unproductive land. Because of corn's limited usefulness as fuel, it is imperative to find other sources for ethanol if it is to be a viable ongoing solution.

Biodiesel is an alternative fuel for diesel-fuel vehicles. It can be distributed through the existing infrastructure, too. Biofuels can be produced from biomass, vegetable oil, and other organic substances such as algae grown in algae farms. Research is being done to find the best feedstocks, or raw materials.

E85 is used in the calculations presented here because it is the most prevalent alternative fuel used today. It also simplifies the projections and provides a conservative estimate since some petrol gas is still used. E85 is a gas/ethanol mixture with 85 percent ethanol and 15 percent gas. The reason this fuel is attractive is that there is minimal modification required to implement the transition. Existing gas pumps can distribute it, and many gas stations already make it available. E85 is less efficient than gasoline but costs approximately forty-five cents less per gallon, and that difference will continue to increase as oil prices rise.

You must have a flex-fuel vehicle to use gas/ethanol mixtures or, possibly, other forms of alternative fuel. Flex-fuel vehicles cost only about one hundred dollars more than petrol gas vehicles. There are 5 million to 6 million vehicles on the road today, starting with cars produced as early as 1999, that can use gas/ethanol mixtures.[8] Check to see if your vehicle is flex-fuel ready and, if not, what it would take to change it.

It takes energy to grow crops and convert them to usable fuel. Ethanol produces only 25 percent more energy than is consumed to grow it, process it, and ship it. Soybean biodiesel returns 93 percent more energy than is used. Researchers continue to improve on a variety of alternative fuels.

Again, I am not advocating any particular solution but simply including alternative fuels as one of the components of a solution. If other alternative fuels are used, the amount of petrol gas will be reduced further. Phasing in fully electric vehicles would, of course, reduce the use of petrol gas even more, which is not reflected in these calculations. The use of a variety of plug-in hybrids will also introduce complexity that is not reflected in these calculations. But this gives us a starting point for a solution.

Now we can combine the changeover to plug-in vehicles with the ramp-up of the use of alternative fuels. In the projections shown in figure 2-6, by 2021, 50 percent of the automobiles that can use petrol gas will also be capable of using alternative fuels instead. By 2035, 100 percent will have that capability.

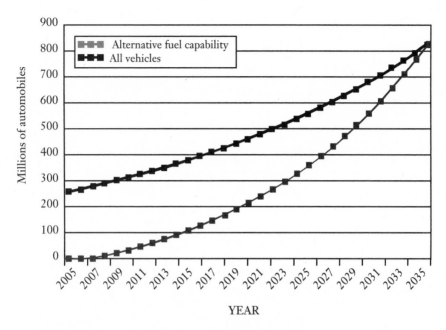

FIGURE 2-6: U.S. Automobiles Capable of Using Alternative Fuels

Now let's put the use of alternative fuels together with the use of alternative vehicles and see how this affects the amount of petrol gas we need.

RESULTING REDUCTION IN PETROL GAS CONSUMPTION   In the second scenario, 100 percent of new automobiles are gas/electric hybrids by 2015. The third option shows plug-in vehicles introduced in 2008. By 2017, 100 percent of automobile production is projected to involve plug-in vehicles. Gas/electric non–plug-in hybrid production is phased out, but previous vehicles can be converted by adding plug-in capability. The final projections (as shown in figure 2-7) include the use of alternative fuel in addition to plug-in hybrids.

As these graphs show, the gas/electric plug-in hybrid, combined with the alternative fuel, eliminates more quickly the need for petrol gas. Since half of oil is used for petrol gas (as figure 2-2 showed), these changes eventually cut the need for oil almost in half. There will still be an oil industry, but having these alternatives will ensure that it won't

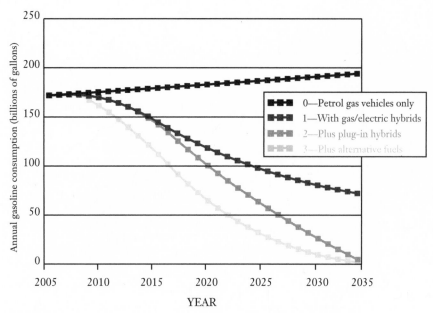

FIGURE 2-7: Gasoline Required to Fuel All Automobiles in the United States

have the potential to crash our economy if the supply dries up or is disrupted in any way.

These options can almost move us past our dependence on foreign oil, since sixty percent of our oil is imported. But now cars will be plugged in at night, increasing the demand for electricity. How will this demand be handled? There are concerns that if the additional electricity comes from coal and other emissions-producing sources, the benefits will be canceled out. However, the use of clean renewables *can* reduce the harmful effects of increased electricity requirements. We'll get to that soon.

CONSUMER FUEL COSTS How does all of this affect the individual consumer? Let's consider a typical consumer, Pat, who has two children and lives in Pensacola, Florida. Figure 2-8 shows the weekly automobile fuel expenses under the previously described scenarios.

The first scenario (2008) represents the gloom and doom of the status quo—with petrol gas prices rising. Pat is spending $49 per week to fuel the car. If prices continue to rise at the rate they have over the

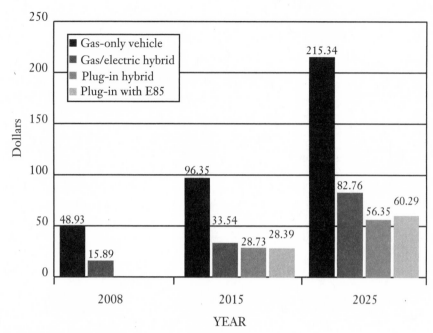

FIGURE 2-8: Pat's Weekly Automobile Fuel Expense

previous three years, the household fuel expenditure doubles by 2015 (scenario two) to $96 per week.[9] No more taking the kids to the movies. Their clothes become a bit worn, but they'll do. Savings? Retirement funds? You've got to be kidding. Pat is really hurting, because the price of oil affects the cost of virtually every product. The economy slides further into a recession.

In the third scenario (2025), Pat's car sits in the driveway. It needs repairs, but the family just can't afford it. Dinner is macaroni and cheese, ramen noodles, or pot pies. Even ice cream for the kids is out of the question. Truth is, Pat has had to decide between getting petrol gas to go to work and buying food for the kids. That's because petrol gas costs $215 per week—four times as much as in 2008 and more than $11,000 per year—not to mention the impact of oil prices on other goods and services.

What a miserable story! For Pat this may be a scary possibility, but many in the world are already in this position. With a comprehensive energy solution, we can create much better circumstances.

The other transportation scenarios paint a different picture. With the use of hybrid vehicles and alternative fuels, the cost to the consumer actually goes down by $10 per week in 2015—a much better scenario! This shows that the proposed plan can get us through the shorter-term problems—if we get moving quickly on implementation. However, even with the hybrids and alternative fuels, costs eventually rise to $60 in 2025 instead of the starting point of $49. Additional improvement is required.

The eventual rise in fuel costs even with all of the proposed changes shows that more needs to be done. We need to make the shift to fully electric vehicles—and/or something else, such as hydrogen vehicles—by 2025. This is entirely feasible. Again, it's imperative we get moving now to make this happen.

Hydrogen Vehicles

The transportation solutions we've focused on so far address petrol gas and work well for smaller vehicles that are mostly used locally: cars, SUVs, vans, light trucks. What about larger vehicles that regularly travel cross-country, such as those that transport food and other goods? Plug-in vehicles take a relatively long time to recharge, so they're not the best choice for long-distance transport.

The second biggest use of oil is for distillate fuel, which includes fuels used in large trucks and buses. The third use is for jet fuel. Addressing these needs can also help reduce our dependence on oil.

Hydrogen, a source of clean, storable, and renewable energy, can be replenished in a matter of minutes rather than hours. Component #3 of our solution is hydrogen for cross-country trucks and buses in the short term, with the possibility of moving to smaller vehicles in the longer term.

The most commercially viable hydrogen technology today exists for larger vehicles like buses and trucks. Hydrogen generally exists in a gaseous form. Gases are not as dense as liquids. If you had a can filled with petrol gas and another similar can filled with hydrogen, the amount of energy you would get out of the petrol gas is much more than you would get out of hydrogen. This is called its *energy density*. The challenge with hydrogen, then, is that the tanks required to replace existing gasoline tanks would be prohibitively large.

If hydrogen is converted to a liquid, the containers to transport it can be much smaller. Liquefied hydrogen has much better energy density than in its gaseous form. However, liquid hydrogen is highly combustible and has special safety and security considerations. These must be addressed.

Mileage for hydrogen vehicles is now one mile per kWh, which is equivalent to 10 cents per mile—the same as \$2.50-per-gallon gas prices. That means it is already much cheaper than petrol gas at hydrogen's current prices. As time goes on, petrol gas prices will continue to rise and hydrogen prices will fall, so the relative cost of hydrogen will get even better.

There are large-scale economic benefits to using hydrogen to fuel large vehicles. Since these vehicles are used to transport goods, a cost-effective transition to hydrogen would ensure that the price of those goods would not be affected by rising motor-fuel costs. This affects all economic groups, not just those that drive cars, and could make the biggest difference in our overall economy.

There are multiple ways hydrogen can be accessed and used to fuel vehicles. Hydrogen is one of the most abundant elements in the universe, but not in its pure form. Therefore, the first consideration is how to get hydrogen into the form needed to provide energy. One method is called electrolysis, which involves running an electrical current through water (H_2O) and separating the hydrogen from the oxygen. Another method is called reforming, which involves a chemical process to separate hydrogen from other substances, such as fossil fuels and biomass.

One of the ways to use hydrogen in a vehicle is in an internal-combustion engine where hydrogen is combusted or burned using essentially the same method as petrol gas vehicles. Hydrogen burns easily, but the combustion itself can cause nitrous oxide—a greenhouse gas we will address in chapter 3 ("Global Warming"). As with petrol gas engines, a catalytic converter or other method is required to take care of the emissions.

Another method of providing energy for the vehicle is using a fuel cell. Fuel cell vehicles propel the car using an electric motor, like a fully electric vehicle, but rather than supplying the electricity from a charged battery, the electricity is supplied through a chemical process between hydrogen and the oxygen from the air. We will cover fuel cells more in the chapter on energy storage.

Hydrogen can be transported in the vehicle in tanks. Hydrogen is highly combustible, so safety is a significant consideration. Alternatively, water could be transported in the vehicle. The hydrogen could be separated out on board via electrolysis, and once the hydrogen is used to power the vehicle, it would combine with oxygen in the air to make water again, forming a complete cycle. With this option, however, the amount of energy you get out of the hydrogen is not much more than the energy required to put into the process. Research is in progress to improve this.[10]

It is a logical progression for oil companies to invest in these new technologies because they already use hydrogen in their plants, and they have experience with distillation processes that can be used in hydrogen processing. However, just as it is illogical for auto manufacturers to shift to electric cars without outside pressure, it is also illogical for oil companies to move away from their bread-and-butter products without outside pressure. The unavailability and inaccessibility of oil is, naturally, one incentive for change. In addition, governmental pressure, consumer pressure, and the introduction of electric vehicles provide gas and oil companies the incentive to work with car companies to move as quickly as possible to hydrogen technology.

Infrastructure

All three of the changes described—plug-in vehicles, alternative fuels, and hydrogen vehicles—would require additional elements to support a transition. A benefit to this is that establishing these additional infrastructures also creates new job opportunities in addition to those created by these new products themselves.

Plug-in vehicles are capable of recharging from a standard 110-volt plug. The additional infrastructure needed for plug-in vehicles includes new electrical outlets in residential garages. Banks of electrical outlets can also be added to parking garages and parking lots at which both customers and employees could recharge their cars.

To support the move to alternative fuels for our vehicles, pumps at filling stations must be converted to supply the new fuels. Converting to hydrogen fuel is a bigger challenge. Currently, there is no infrastructure to support it—there aren't many places where you can get refills on your hydrogen. A completely different and more complex infrastructure would be needed. For example, the proposal by our capstone team included hydrogen tanks similar to those that supply gas grills with propane, which is also combustible.[11] The tanks could be exchanged at hydrogen filling stations. Equipment to load and unload the tanks would have to be installed, vehicles would have to be designed to accommodate the exchangeable tanks, and the tanks themselves would need to be engineered.

There may be many hydrogen solutions, but any solution would need to consider the technical means to access the hydrogen, the physical means of carrying it in a moving vehicle, and the distribution system to get it to the vehicle. Iceland is considered a model of what is possible. Filling stations and hydrogen buses are already in operation throughout that country.

There is a catch-22 with the transition to hydrogen vehicles: Until there are enough hydrogen vehicles, gas stations won't be motivated to install hydrogen-refueling stations. Until there are enough hydrogen-refueling stations, automakers won't be motivated to develop hydrogen cars.

Something is needed to stimulate adoption of hydrogen-refueling stations, and the process needs to be financed. Estimates of the costs of the hydrogen infrastructure alone are in the range of $15 billion to $19 billion.

The winning team in our capstone competition came up with the idea of a hydrogen initial public offering, or IPO.[12] The term is usually used to indicate that it is the first time that a company would offer their stock for purchase by the public. In this case, it means that it is the first time that the government would offer licenses to operate hydrogen-fueling stations via an auction. This method was used to designate radio frequencies for radio stations and cellular spectrums for cellular-phone companies.

The number of licenses per geographic region would be limited to create competition. Anyone could bid. This approach would not only provide an avenue for gasoline-entrenched businesses to transition to a hydrogen economy but also provide incentives for small-business participation. Since the government would be auctioning these rights, this approach would create the public funding for the hydrogen infrastructure.

To make their idea work, it would be best to initially focus the infrastructure transition on installing hydrogen stations along major highways, since hydrogen fuel would initially be used by cross-country buses and trucks. Once the technology was further improved, the focus could move to hydrogen cars and developing a hydrogen infrastructure in neighborhoods. The viability of this path would be determined by the market for vehicles at that point.

It is a common belief that gas and oil companies will absolutely oppose all attempts to move to other fuel sources. However, since they get profits from the products sold at their gas stations, a hydrogen solution gives the oil and gas companies a way to maintain their profit-producing stations.

As the remaining sources of gas and oil become harder to access, they will become more costly to extract. If there are no alternatives, those costs translate into higher prices to the consumer, and can provide record profits for gas and oil companies. This is exactly what has been occurring. But if there are cheaper alternatives available, consumers will naturally turn to those alternatives, negatively affecting oil company profits.

As oil prices rise, it is only a matter of time before alternatives become financially competitive. Plug-in hybrids and fully electric cars give the gas and oil companies motivation to move to producing and distributing hydrogen fuel as quickly as possible, and the markets will determine when and where hydrogen-fueled vehicles are an appropriate solution.

U.S. oil companies will still be producing the less refined products including asphalt, plastics, etc. Since these are less costly to produce, this may be a benefit toward maintaining good profits when oil is extracted from lesser quality sources.

Other Transport Options

When we consider the amount of energy used for transportation, we can assess it per mile traveled, but what if more people travel in the same vehicle? If ten units of transportation fuel is consumed and there are two people in the vehicle, then that's only five units of fuel per person instead of ten. Therefore, we can reduce the need for fuel by traveling together when we're traveling in the same direction—carpooling is one way to cut the fuel demand.

If ten people are in the same vehicle, that's only one unit of fuel per person. Of course, larger vehicles such as buses use more fuel per mile traveled than a car uses. However, a bus that consumes ten times as much fuel as a car, but carries twenty people, is consuming half as much energy for the service it provides.

Of course, this will be true only as long as the planes, trains, and buses are full. From a traveler's standpoint, it can be much more productive and enjoyable to be a passenger rather than have the stress and frustration of being a driver in busy traffic. Passengers can read, sleep, work, or socialize. Some who commute on a regular basis organize card games and other activities.

Reducing Transportation Needs

We will need less oil for transportation if we cut down on travel itself. This includes reducing unnecessary travel, whether it be local or long distance.

Businesses can consider using remote communication when appropriate. Individuals might choose to vacation closer to home. We can walk or bike for errands and local travel, using our personal energy as fuel for transportation rather than fossil fuels. The added bonus of exercise can make our own "body vehicles" operate more efficiently.

Each person needs to make an individual decision on when these actions are appropriate. As I said at the very beginning of this book, energy management also applies to people. Sometimes more human energy is required to get a job done. Communicating in person can provide more energy than communicating via a teleconference or video, which, again, can provide more energy than does written or verbal communication. Business travel will still be necessary, depending on the importance of the work at hand.

Vacations will also still be necessary to replenish our personal energy and to keep us connected to people, nature, and our planet. Again, a physical trip will have more effect than a video or photo, which will have more than a letter or phone call.

The key is to consciously choose. Instead of traveling just because it's the thing to do with vacation time, or for business, we can decide whether it's what we really want to do, or what we should do. We can make an effort to look for opportunities to reduce our transportation demands. Not only is it good for the planet; it also saves us money. And as we shift to these alternative-energy sources for transportation, we are all better off.

Buy Locally

Besides reducing the amount of oil required to move us around, we can reduce our dependence on oil by decreasing the amount of transportation required for the things we buy. That means buying products that have been grown or produced locally, which is good for the economy of your community.

We can also cut down on the transportation of the energy itself. When we address the matter of energy supply, you will see how to choose sources based on their physical availability. For instance, biomass is especially good for farm machinery because sources of biomass—crops and animal waste—can be processed nearby and then turned back into fuel for the machinery.

It is true that if 100 percent of our large and small vehicles are electric and/or hydrogen fueled backed by clean, renewable energy, our dependence on petrol oil and the emissions caused by transportation are totally eliminated. But, if we establish more energy-efficient practices and habits now, they will still serve us well for the long term.

OTHER OIL-BASED PRODUCTS

As figure 2-2 showed, petrol gas accounts for almost 50 percent of the oil we use. As we've already mentioned, alternative fuels can be used to replace diesel fuel, which is currently used in transport trucks, farm machinery, and some passenger vehicles. In addition, reducing unnecessary air travel will reduce the need for jet fuel, and converting airplanes to hydrogen or alternative fuel will reduce the oil demand even more.

There are many other ways that we, as consumers, can reduce our dependence on oil as well. We can all reduce our consumption of plastic, which is an oil by-product. Using cloth shopping bags instead of plastic bags, reusing plastic water bottles, and recycling all plastics that are recyclable will help reduce the consumption of oil.

In addition, we can stop mindlessly accepting many toys and trinkets made of plastic, many of which a child never plays with. Be sure the child wants it and will play with it before accepting plastic freebies at stores or buying plastic gifts. It is up to each person to decide where and how to make changes. The important thing is to start.

3

GLOBAL WARMING

IS THERE GLOBAL WARMING or not? Is it caused by humans or by natural phenomena? Is that really the point? I finally thought, who cares? Rather than wasting our human energy debating this, why not use that energy to find ways to be more efficient with the Earth's energy? Why not find ways to reduce emissions so that we have good air quality, regardless of how our current problems were created?

I also realized that reducing energy use and our dependence on oil wasn't the only requirement to reducing global warming and emissions. I can use less energy, but if it still produces emissions, there is more work to be done . . .

EMISSIONS

Oil, natural gas, and coal are all fossil fuels, which means they are created from the fossilized remains of plants and animals, including those that lived millions of years ago. In other words, the remains of dinosaurs are fueling our cars!

When we drive our cars, heat our homes and businesses, and use electricity, we burn these fossil fuels, and when they burn, gases are emitted into our atmosphere. The gases include carbon dioxide, or CO_2. CO_2 is the most prevalent of the gases produced by burning fossil fuels.

Greenhouse Gases

CO_2 is called a greenhouse gas. Other examples are methane, nitrous oxide, ozone, and man-made gases such as aerosol sprays and fluorocarbons, which are used in air-conditioning and refrigeration, as well as water vapor.[1] CO_2 makes up 77 percent of the greenhouse gases. Humans themselves emit CO_2, methane, and nitrous oxide.

The term *greenhouse gases* refers to the tendency of these gases to create a bubble effect around our planet. When greenhouse gases stay in our atmosphere, they prevent the Sun's heat from leaving our atmosphere, trapping it, similar to the way greenhouses retain heat to warm plants.

This greenhouse effect is one of the factors believed to cause global warming. Global warming is the increase in the average temperature near the Earth's surface. Weather patterns change, almost as if the Earth were sweating, or shivering with fever, leading to hurricanes, increased rain, drought, increased volcanic activity, and tornadoes. Species of plants and animals, including humans, are affected as well, as each species is designed to operate best under certain climate conditions.

With heat trapped in the Earth's atmosphere the polar ice caps melt, making the oceans colder, just as it would make your bathwater colder if you added large amounts of ice to it. With your bathwater, the water would stay colder, but since more heat is being applied from the atmosphere to the oceans, some of that heat counteracts the cooling effect of the ice on the ocean water.

Ice and snow reflect heat better than water and land do. When ice and snow disappear, heat that used to be reflected is absorbed, further contributing to warming. In addition, ice and snow that run off from the land into the oceans cause the sea level to rise, threatening large numbers of coastal cities.

Another effect of the increase in temperature is that long-frozen soil thaws out. Methane that has been trapped in that soil is released, adding to the greenhouse gases and further aggravating the problem. This creates urgency to stop the thawing as quickly as possible.

Since burning fossil fuels produces emissions that can affect global warming, a way to reduce global warming is to reduce emissions to a level that is in balance with the Earth's atmosphere. And again, whether or not we are creating global warming, these emissions also affect air quality. Since air is the primary element our species needs to survive, it makes sense to minimize harmful emissions so we have the best air quality possible.

Toxicity

Toxicity is a measure of how poisonous something is. Any chemical can be toxic when used in the wrong context or if present at a certain dosage—the "too much of a good thing" phenomenon. As we saw previously, water is one of the culprits in creating greenhouse gases, despite being vital to our survival. If we get too much water in our lungs, we drown. Water isn't toxic, but too much water is. Toxicity isn't an absolute condition; it's relative to the situation.

Ideally, a system such as our bodies or the Earth will have balanced inputs, internal processes, and outputs so that they can maintain dynamic equilibrium. That means the system is stable, appearing on the outside to stay the same, but when you look more closely, you will find that it is actually constantly changing. For instance, we breathe in and we breathe out. If either of these processes stopped, we would be in trouble. In another example, cells in our bodies are constantly dying, while others are being born. These two processes are critical to our survival. As long as they stay in balance, we stay healthy.

When we talk about something being toxic, we generally mean that it is toxic to us. But if the circle of life is in balance, those things that are toxic to us are valuable components for something else. For instance, dead animal carcasses produce smells that are unpleasant to us and that would definitely be toxic if we ate them. However, for buzzards, they're food.

In terms of atmospheric composition, certain levels of CO_2 are toxic to us. However, they aren't toxic to plants. As a matter of fact, CO_2 is a vital component of photosynthesis, in which plants use the energy in sunlight to convert CO_2 and water into oxygen.

CO_2 is vital for the survival of plants. We produce CO_2 as our waste product. Oxygen is vital to our survival. Plants produce oxygen as their waste product. We have a reciprocal relationship with plants that is critical for the survival of both. So CO_2 is good in the sense that it feeds plants, which then feed us.

A certain amount of CO_2 is healthy for the Earth, but too much isn't good. Therefore, we need to find a way to reduce the emissions to an acceptable level and/or find a use for these gases that produce good effects elsewhere. We need to convert CO_2, or anything for that matter, to what we need when we need it, and then convert it to something else when that is needed.

Population

I have indirectly addressed two reasons why the population affects global warming. The first is the fact that as the number of people increases, so does the demand for energy. Energy demand has traditionally been fulfilled by fossil fuels, which produce emissions. Therefore, the population has had an effect on emissions and will continue to do so unless we find ways to supply energy by other means.

In addition, each person's body produces CO_2 emissions. We breathe out CO_2. The daily output of CO_2 by the average human is around one kilogram. It's pretty obvious, then, that as the population grows, emissions go up.

The birth rate and death rate contribute to the population size. As we find ways to live longer, the total population increases. As the birth rate goes up, the population increases. As the population goes up, the number of people available to have children goes up, which can also cause the population to increase. All mammals—cows, horses, dogs, cats, pigs, etc.—breathe out CO_2. The sheer number of mammals, therefore, is also a consideration in CO_2 emissions.

We need to be aware of these factors so that we don't overrun the Earth to an unsustainable level.

Carbon Footprint

Our lifestyles determine the amount of carbon each of us generates. The term *carbon footprint* is used to indicate the amount of carbon we produce as we walk through our daily lives.

There are many online calculators that will help you determine how much carbon emissions result from your lifestyle. They ask questions about the vehicle you drive, how far you commute, how often you travel, how you travel, and the energy used in your home. They ask about other lifestyle choices as well, including the types of food you eat, whether you recycle, and elements of your buying habits.

Once you've measured your carbon footprint, you can determine ways to reduce it. You can take actions we've covered and others that are yet to come. In addition, continue to look for opportunities by thinking of the categories that are covered in this book and coming up with your own solutions within the categories.

OFFSETTING EMISSIONS

As I've said before, the problem with carbon dioxide is not the amount that is emitted but rather how much is left in the atmosphere. In addition to taking action to reduce carbon emissions, we can counterbalance, or offset, the emissions.

Reforestation

We can take advantage of a very natural process to reduce carbon dioxide. Just as we breathe in oxygen and breathe out carbon dioxide, plants do the reverse—they "breathe in" carbon dioxide and "breathe out" oxygen. Therefore, one way to counteract the increase in emissions is to increase the number of plants and trees.

Each tree offsets your environmental impact by breathing in about 730 kg of CO_2 emissions over its lifetime. It is estimated that the average person needs to save about 7,000 kg of CO_2 per year, so planting just ten trees each year is one strategy for achieving this.[2]

One mature tree absorbs approximately 13 pounds of carbon dioxide a year. For every ton of wood a forest grows, it removes 1.47 tons of carbon dioxide and replaces it with 1.07 tons of oxygen. One acre of forest absorbs 6 tons of carbon dioxide and puts out 4 tons of oxygen, which we all need every day.

These are factors that need to be taken into consideration when deciding to cut down forests and use wood for products that could be produced from other materials. At the very least, the practice of replenishing what is cut down will help.

That's why planting trees is a common activity for carbon-offset programs. Trees also block the heat of the Sun in the summer and provide shelter from the cold wind in winter. A bonus is that they're also pretty!

Carbon Offset Programs

After you have done everything you can to reduce your own emissions, you can help counteract the remainder by buying carbon offsets. For instance, you may have the desire to have ten trees planted a year to offset your personal CO_2 emissions, but the reality is that you don't have the location or the inclination to do the planting yourself. Instead, you can use carbon offsets—programs to which you contribute money to support projects such as tree planting, providing energy-efficient products to the poor, and investing in clean energy.

Donating your money to local projects enables you to see the results for yourself. You can ensure that your money is used appropriately. You'll also be benefiting your own community, thus benefiting yourself! Trees that are planted locally will produce oxygen in your vicinity, plus you get to enjoy the aesthetic value. If carbon offsets are used to buy light bulbs for the poor in your community, the recipients benefit by reducing their energy costs and you benefit by reducing the possibility of hitting peak electricity demand in your area, which will be covered in a later chapter. Of course, don't stop at the local level; it's just a good place to start.

It's important to recognize that offsetting your own carbon production through a carbon offset program is not an alternative to reducing your own carbon footprint! If you would like to contribute more than the amount needed to offset your emissions, that is obviously welcomed.

Buy "Green"

Many companies are actively seeking to reduce their emissions. You can keep informed about companies that are "going green" so that you can indirectly reduce energy consumption and emissions through them. If you order a computer or other product or service from a company that creates less emissions in producing that computer than another company creates producing theirs, you have reduced total emissions via your purchasing choices.

For the companies themselves, there is a marketing benefit to undertaking green initiatives. The public image of the company is improved by these initiatives, giving them an edge over their competition. The efforts to reduce emissions thus can provide benefits to the company's income level, to their energy-related expenses, to the air quality of their community, and to the planet. As more and more people place an emphasis on buying from "green" companies, it will provide the incentive for other companies to reduce their own emissions.

WHAT'S POSSIBLE

Working together, we can accomplish more than we may believe is possible. The chart below shows how synergy works—how putting multiple components together can add up to more than just the sum of its parts. In this book we are covering four main areas that can be improved on. Let's look at how these four areas combine to reduce emissions and global warming.

	Consumers (businesses and individuals)	**Suppliers**
Energy Usage	Energy-efficient products and practices	Efficient storage and distribution
Emissions	Clean-energy products	Clean-energy generation

Let's say that consumers focus on reducing their energy consumption. That means both businesses and households turning off lights when they're not in use, buying energy-efficient bulbs, turning thermostats up or down to minimize energy use, etc. It might mean taking baths and showers less often, cutting lawns less often. However it's achieved, though, every household and business would set a goal of reducing their energy consumption by 20 percent.

Secondarily, let's say that consumers focus on reducing their emissions. That means buying products that have lower emissions, such as electric lawn mowers, solar energy products, low-emissions cars, alternative fuels, etc. Let's give them a goal of a 20 percent reduction in emissions.

Now consider the energy suppliers, which we'll get to soon. Let's look at their emissions. We'll give all energy suppliers the goal of reducing emissions by 20 percent through the use of clean energy.

The remaining component is reducing the amount of energy that energy suppliers must produce to meet the demand. This means improving efficiency and reducing waste. Energy is wasted if it is produced but not used. This is where the global grids, smart grids, and energy storage come into play, which we will soon cover. While we haven't estimated available energy in other countries, a global distribution grid helps minimize the impact of the aggressive ramp-up in demand in countries such as China and India. We'll set a goal of 20 percent for energy suppliers to increase efficiency.

If we reduce each of these four components by just 20 percent, that means we could cut emissions by more than half. To understand this, consider this example. Say you go to a store where they have one hundred marbles. You buy 80 percent of them. That's eighty marbles. When you take them home, you notice that 20 percent of them are flawed in some way, so you keep only 80 percent of them. That means sixty-four are good. Now let's say 20 percent of the remaining marbles are an unappealing color, and you decide not to use them. So you use 80 percent of the sixty-four marbles, which is fifty-one. But then you discover the box you want to store them in can hold only 80 percent of the remaining marbles, leaving you with forty-one.

In the same way, if we combine these four areas of improvement (80 percent of the current consumer energy usage × 80 percent of the emissions coming from the products that consumers use × 80 percent of the emissions as compared to the emissions from current energy production sources × 80 percent due to efficiency gains in production, storage, and distribution) we can get to .80 × .80 × .80 × .80 = .4096, or 41 percent of the current emissions we're producing. Stated differently, these improvements would reduce emissions by almost 60 percent, or more than half of what we would otherwise produce.

4

ENERGY DEMAND

ONE DAY, MY SON came out of his bathroom with a very annoyed look on his face. A light had burned out, causing him to look up at the light fixture. "Mom, there are those old kind of light bulbs in my bathroom!" I sent him on a search for others. I was surprised that he found a dozen of them. It wasn't just because we had overlooked them, though. We were uncertain whether we should wait until they burned out before replacing them. Now we've done the research and we know better.

The amount of available information on energy topics is overwhelming. It also seems conflicting sometimes. Different lifestyles, different climates, different rating systems for different products—all are factors in making the right decision for a given situation.

I have these fits of guilt, wondering if I'm doing enough, wondering if there is some new discovery that I haven't incorporated into my lifestyle yet. After all, I don't think about global energy all of the time. And new products, services, and information are coming out all of the time.

I decided I needed some trigger to get me to think periodically about these things. Paying the utility bill—that's it! I can keep articles and look monthly to see what additional steps we can take to do our part.

GREEN LIVING

Energy management involves balancing the energy supply and the energy demand because if we reduce the amount of energy we consume, then we can reduce the amount that needs to be produced. Therefore, an absolutely critical element is reducing the demand.

A large part of reducing our energy demand can be accomplished by simply putting some attention to it—turning off lights, changing temperature settings, combining multiple tasks into one trip. Some are healthy—walking, biking, hanging clothes out to dry. Some promote family togetherness. Some increase a feeling of community—sharing rides, sharing other resources, working together on projects, buying local products to reduce the energy used in transporting goods.

Instead of remembering the traditional three Rs—reading, (w)riting, and (a)rithmetic—we can focus on the three Rs of energy-demand reduction: reduce, reuse, and recycle. By doing these three things, we reduce the energy needed to extract, produce, or otherwise process resources.

That doesn't mean you're cheap, though it will save money. It can become a fun experiment to see how much you can reuse, what you can fix, and what creative uses you can find for things. And you're doing something good for yourself as well as the rest of the planet!

It's worth thinking of these three Rs in sequence.

Reduce

First, think about how to reduce—reducing waste and energy usage. There are many obvious actions to take to reduce energy use: turning off lights, using less heating and/or air-conditioning, using less fuel by walking, biking, carpooling, and taking public transportation.

You can reduce waste by limiting the paper and plastics you use, including disposable coffee cups, plastic water bottles, paper napkins, and paper towels.

Using less water reduces the amount of energy needed to treat and distribute it. Therefore, watering lawns less often or replacing grass with groundcover or other landscaping that uses less water—or even skipping showers—can make a difference in energy consumption. Some people would love an excuse to do that!

Reducing the number of meaningless trinkets, unnecessary print-outs, and any type of production waste reduces the energy needed to produce and transport.

Reducing the amount of food consumed reduces the energy needed to grow it, process it, and distribute it, as well as the personal energy needed to carry it around.

Look for opportunities besides those provided throughout this book. You'll be amazed how many you find.

Reuse

Second, consider how you can reuse something or how someone else may be able to reuse something you no longer want or need.

Print drafts of documents on the back sides of paper. Find a new use for a piece of clothing. Donate items that are no longer used, sell them in garage sales or online, or give them to friends, thus enabling the items to be reused by someone else. Reuse plastic bottles as well as shopping bags, both paper and plastic. Better yet, reuse cloth shopping bags to reduce the need for paper or plastic bags.

More opportunities for reusing materials are covered in chapter 5 ("Green Building").

Recycle

Third, consider how to recycle items. Recycling means an item will be passed through a series of processes so that it can be used again in a new form. Since it requires energy to process, this is the last choice to consider, but it can make a considerable difference.

Recycled newspaper and other paper can be used in recycled paper products or for fireplace logs, pencils, and even decorative items. One four-foot stack of newspaper is equivalent to one forty-foot fir tree. And reducing the number of trees that are cut down has a double benefit of offsetting emissions, which we covered when we addressed reforestation.

Recycling services are available in most areas for glass, plastics, paper, and metal. Old computers, cars, appliances, and other machines can be recycled so their materials can be used again. You can recycle plastic bags by returning them to stores. Enough with the bags already—but as we've seen, they are made from oil, so you're making a difference in two ways!

Another excellent way to recycle is to compost nonanimal kitchen waste. When the materials break down, they recycle into fertilizer. Homes can compost, and so can restaurants, hotels, and company cafeterias.

None of these actions mean you have to deprive yourself or inconvenience yourself. It means being aware so you can make more informed decisions. It may also mean looking at the entire picture.

For instance, walking or biking uses your personal energy. It takes longer, and time may be a critical resource for you. The weather can enter into the decision. So will the distance. It may or may not be the right choice, or it may work only on certain days.

On the other hand, using your personal energy to transport yourself can have environmental benefits and personal benefits. The exercise can create a stronger, fitter body and increase the efficiency of your body.

Alternatively, you can use public transportation. It takes less of your personal energy since someone else is doing the driving. You can read, relax, or get work done while you're traveling.

Using your personal energy or sharing the energy expenditure with others reduces the overall petrol gas, electricity, or other energy source required. And it feels good to do good! The difference is in considering energy use and waste in your decisions.

The key is for each individual or business to truly decide the best decision in each situation. There isn't a "one size fits all" answer. But the impact of all of the actions you take to reduce, reuse, and recycle is significant. And, again, it can be fun to see how creative you can be in finding new uses for things and ways to conserve.

Water

Water is the second most critical element needed for human survival, second only to air. It is not infinitely available. And, as with other resources, it takes energy to treat and pump the water to us. The water utilities are generally the largest users of electricity. Therefore, it is imperative that we be conscious of using less water and looking for ways to be smarter about how we use it.

Xeriscaping (that is, using native landscaping) reduces the need for watering lawns and gardens. Xeriscaping does not mean you have a yard full of cactus plants. It means using plants that are natural to the climate. It means less lawn area, and arranging plants to minimize the need for water. While you're at it, you might think about planting a garden to grow crops. You can then share the fruits (and vegetables!) of your labors if you have excess.

We can reuse water, too. Water that contains human waste is called blackwater, which needs to be treated before it is used for irrigation. Water from washing machines, baths, and dishwashing is called graywater. We can recycle graywater by using it directly for irrigation, which reduces the need for treating and pumping it. First check your local standards to see if there are any restrictions due to the detergents used or if there would be drainage into any local bodies of water.

Rainwater collection systems provide an excellent alternative source for water. You can purchase a rainwater collection system or make your own.

Low-flow toilets and low-flow showerheads also reduce the amount of water used. Even simple actions like taking shorter showers and turning off the water while you brush your teeth can make a difference.

PRODUCT EFFICIENCY

Demand is also affected by the efficiency of the products we use. Efficiency essentially means we get the same results while using less energy. Our own personal efficiency means getting the same result while using less energy.

Some products and energy-related processes lead to energy loss—energy is wasted. For instance, when there is resistance in an electrical circuit, energy can be lost due to dissipation of heat. "Heated arguments" are a sign of resistance, too, and an indication that human energy is being wasted! Rather than arguing about problems, let's use that energy for solutions to inefficient use of the Earth's resources.

Products

If you're wondering which products can make the biggest difference in your energy bill, think about how they are used. Products that we use most often and that consume the most energy when we use them will make the biggest difference in reducing energy use. Since refrigerators run twenty-four hours a day, they are usually the most significant appliance in the home to consider for energy efficiency.

While washers, dryers, dishwashers, stoves, ovens, and microwaves are not used constantly like a refrigerator, they draw a lot of energy when we do use them. We can cut down on energy consumption by using energy-efficient units.

Products such as TVs, ceiling fans, and home computers may not use as much energy as appliances, but we have a tendency to leave them on even when they aren't being used. VCRs, stereos, coffee machines, and other smaller products still make a difference. By buying efficient products, and turning them off when not in use, we are making a difference in our energy consumption.

Many appliances use energy even when they are turned off. Anything that has a timer or a clock on it or that is kept in ready mode so that it can power up quickly is using energy in its standby mode. These need to be turned off at the light switch or unplugged to cut their energy use. Alternatively, there are power strips that can be used to turn these off completely.

Our personal habits and practices make a difference, too. Dishwashers, washing machines, and dryers can be used on low or no-heat settings. We can check that they are fully loaded before we do run them. Water heaters and refrigerators can be set to optimum settings rather than maximum settings. We can adjust our habits to turn off all energy-consuming devices when they are not in use.

Light Bulbs

This is a good time to talk about light bulbs. Believe it or not, if we all changed our light bulbs, it could make the difference of 20 percent reduction in energy usage.

Count the number of light bulbs in the room you're sitting in right now. It's probably at least three. Count the number of rooms in your home. The minimum is probably four—a kitchen, living room, bedroom, and bathroom. That means a minimum of twelve light bulbs. But if you live in a 1,500-square-foot, three-bedroom, two-bath home, it's likely that you use twenty-five or more light bulbs. Larger homes use even more, and offices, retail centers, manufacturing sites, and other business areas use still more.

You might argue that you turn on only one light at a time. You'll probably find that's not the case.

You might think that it's best to wait until your existing light bulbs burn out before you replace them. After all, it's wasteful to throw out something that still has useful life, isn't it? With light bulbs, quite simply, the answer is no.

First, look at the energy savings. Compact fluorescent lights, or CFLs, can save you about thirty dollars in energy costs during the life of that light bulb, as compared with the energy used for a regular incandescent bulb—the ones that have been used for generations. Each CFL costs about two dollars. So you save twenty-eight dollars—per light bulb!

Now look at the cost of the light bulbs themselves. Currently, CFLs cost two dollars to three dollars each and incandescents cost fifty cents. That means CFLs cost four to six times more than standard incandescent bulbs. But CFLs last eight times as long, so you would spend three dollars for CFLs or four dollars for incandescents for the same lighting coverage.

Now look at the impact on emissions. If you save thirty dollars on energy, you've eliminated the need to generate that thirty dollars' worth of energy, thus reducing any emissions from generating it. While the emissions don't directly affect your pocketbook, they do indirectly affect the quality of your life.

What about the waste of throwing out that incandescent light bulb? Doesn't that cancel out the benefits? Nope. Not when we compare it with the benefits of making the change. It is much better to throw out that fifty-cent light bulb to gain the benefits described.

Some people wonder about the light quality of compact fluorescents, as compared with incandescent bulbs. There are some minor differences. For instance, CFLs take a split second longer to come on. When they do, they take another second to warm up fully. But that can be useful for adjusting your eyes to the light, and does two seconds really make that much difference?

You have a choice between soft-white bulbs, bright-white, and daylight. It's really just a matter of preference. Soft whites are the yellowest and work well where you want soft lighting, and daylights are the bluest. Try them out and see what you like. It's really not that tough.

CFLs are rated differently. A thirteen-watt CFL produces the same amount of illumination as a sixty-watt incandescent bulb. That's why they save you money—you're getting the same amount of light but using forty-seven fewer watts. So you do need to make yourself aware of the new ratings.

And don't forget to recycle these light bulbs, too. They have mercury in them, so they are not meant to go into our landfills but should be recycled at hazardous waste centers—the same place you recycle your paints and chemicals.

Another type of lights uses light-emitting diode (LED) technology. These are commonly used as indicator lights on toys and as the small diagnostic lights on computers and other equipment. They are also now available for Christmas tree lights. Larger versions are excellent for task lighting because the light goes directly where you aim it—one of the reasons they're more efficient.

Currently, LEDs are used for special applications because they are more costly, but they are even more energy-efficient than CFLs and generate much less heat than other types of bulbs. They will surely become more affordable over time and will move to more general applications.

So, if you want to make a difference, and save yourself money in the process, go change your light bulbs!

Energy Star Program

How do we know which products are most efficient? The Environmental Protection Agency (EPA) joined with the Department of Energy (DOE) to establish standards for assessing the energy efficiency of products. It is called the Energy Star program. It assesses office equipment, home electronics, and appliances, as well as industrial, commercial, and construction-related products. The Energy Star rating provides information to help you make informed decisions on construction-related products and appliances.

To receive the Energy Star designation, products must meet strict efficiency criteria. The criteria themselves are continually revised so that efficiencies steadily improve. Whenever you consider purchasing a product that uses energy, look for the Energy Star label.[1]

ADDITIONAL BUSINESS CONSIDERATIONS

Business Process Efficiency

In addition to the examples previously described, businesses can have a great impact on energy demand by becoming more efficient. If you process or produce one hundred of something and only eighty are good, there's lots of waste. You end up spending lots of energy and resources on the three Rs that you don't want: redo, rework, and repair. Indirectly, it also affects your energy demand because it means you need more energy to run your equipment, your computers, your offices. Making your business operations more efficient will also help reduce the demand for energy. It's good for the profitability of the business and, you guessed it, for the planet!

Again, being efficient with energy is good on multiple levels. It is just a matter of looking for the opportunities.

Information Technology

Besides the energy required to operate the buildings themselves, businesses have a large electricity demand arising from their computers, printers, and data centers. Some industry analysts predict that energy costs will soon be 50 percent of total information technology costs. There are two ways to combat this. One is to reduce the demand. I'll cover that in this chapter. The other is to reduce the per-unit cost of energy, which I will cover in chapter 7 ("Energy Supply"). There are also other ways to take control of energy costs, which are described in chapter 8 ("Energy Storage").

PERSONAL COMPUTERS   Desktop and laptop computers use energy when they are turned on, whether they are actually in use or not. Get in the habit of turning them off whenever practical.

Computers also generate heat. The miniaturized electronic circuits are primarily responsible for the heat generated. Power supplies generate heat. The faster motors spin, the more heat they generate and the more cooling they require. The faster a disk spins, the more energy it takes to spin it and the more cooling required. Therefore, turning off your computer can also save on the energy required to regulate temperature.

While these considerations may seem small when you consider an individual computer, they can add up to a lot when you factor in the millions of computers used each day.

DATA CENTERS   Companies generally have a central location called a data center where they store data and manage their computer networks. You can imagine it as a whole bunch of electronic file cabinets to house all of the information used by the people working in your office building or for your company or Internet provider.

Data centers are full of heat-generating equipment. The servers plus the tape drives and other motor-powered devices generate heat. Therefore, a room full of computing equipment would be a very warm room if it wasn't cooled. Cooling requires energy.

To counteract the heat produced, one approach is to air-condition the space. Air-conditioning requires electricity. Computer equipment doesn't really care much about the temperature of the air, though; it really cares about keeping its electronic components cool. Therefore, a more energy-efficient option is to use large chillers to cool these areas. Chillers run chilled water through pipes that are located close to the equipment that needs cooling. They cool the data centers, and although they require electricity, they reduce the overall demand for energy.

COMPUTER POWER MANAGEMENT   Another way that data centers reduce energy use is through their function of running the computer networks. It's called computer power management, and it involves shutting off a large organization's computers through the use of special software and methodology when they are not in use.

The concept is similar to the motion-sensitive lights sometimes used for home security, and in public restrooms and other public places. If there is no movement detected for a long period of time, the lights are automatically turned off. These motion-sensitive lights can be regulated so they operate only during certain hours so you don't end up sitting in the dark in the middle of the day! Logic can also be built in to regulate when computer power management is to be used. There are also ways to ensure that work isn't lost if computers are shut down.

Computers already have the capability designed into them to be able to accept these commands from the central data network. As with light bulbs, it may seem that small savings won't make a big difference, but the truth is that lots of little savings add up to a whole bunch.

5

GREEN BUILDING

NOW WE'RE BUILDING A house. We definitely wanted to build "green." But, again, there is so much information out there that I was feeling overwhelmed. Some products are good for warm climates and some for cold; some are good for renovating and some for building new.

We're well aware that a house is a whole system and the pieces have to fit together. What pieces, though? I decided I needed to understand some basic categories before I could even begin to solve this puzzle. I now feel I have a better framework. And it is indeed like putting together a puzzle: mentally challenging and aesthetically pleasing in the end. A green building just feels good!

GREEN BUILDING PROGRAMS

Since more than 70 percent of our electricity demand goes toward running our homes and businesses, construction needs a special focus. The main categories of buildings are industrial, commercial, and residential.

Each type has different proportional uses of energy, but the categories of uses shown here apply to all of them. In this section, we'll cover the different aspects of building. You can decide which aspects you want to implement first, keeping in mind that some of the pieces must be implemented simultaneously to really provide the benefits.

To understand how much energy we use, consider that it takes approximately one megawatt of electricity to power 250 homes in Texas at the same time. A megawatt is an instantaneous measure of demand where a megawatt hour is over a time period: so, it takes 1 million watt hours to power just 250 homes for an hour. Standard incandescent light bulbs come in forty-, sixty-, seventy-five-, and one-hundred-watt sizes. There's a lot of opportunity for reducing energy consumption in addition to light bulbs! We need to consider how to run more energy-efficient and emissions-friendly homes and workplaces on multiple levels.

With more than 100 million households in the United States, a small reduction in energy usage per household can make a huge difference overall. If we include the impact that businesses can make by reducing their energy usage, it is even more significant. The United States currently is responsible for approximately 25 percent of the world's energy consumption but has only 5 percent of the world's population. Each household, each business can make it a project to reduce its energy consumption. It saves money.

New buildings need to be designed for energy efficiency, and existing ones can be modified to incorporate certain elements. There are several aspects of a building's design that specifically affect its energy consumption. First of all, we'll look at heat-related aspects. While I am not advocating particular products or ways of achieving these results, I am advocating the need to address the categories.

There are national and regional programs that promote and reward green-building implementation. The U.S. Green Building Council provides a national standard for rating buildings against their Leadership in Energy and Environmental Design (LEED) criteria. There are four certification levels: certified, silver, gold, and platinum. They correspond to the number of credits awarded in six design categories.

Because the climate makes such a difference in determining smart energy choices, local and regional green-building programs are also very valuable. They can provide specific standards and recommendations to suit your local environment.

These programs are currently voluntary. They raise awareness of green-building standards and practices and provide a means to ensure that claims of building green are valid. They are also a means to get educated on the many elements involved and to ensure that you incorporate all of the necessary components. In turn, you get the desired reduction in energy costs and improvements in quality.

Austin Energy, for instance, has green-building programs that are the most successful utility-sponsored program of their kind. They provide rating systems for commercial, multi-family, and single-family residential homes. Austin Energy also provides substantial support for green-building practices, ranging from rebates for customers who install solar systems to distributing compact fluorescent light bulbs. The utility has also provided free programmable thermostats, low-flow toilets, rebates on hybrid vehicles, and more. Find out what is available in your area.

CREATING INTERNAL TEMPERATURES

A vital consideration in this section is dealing with heat, so we begin with regulating temperatures. It takes energy to raise or lower temperatures. Have you ever walked into a building where it's freezing cold in the summer or stiflingly hot in the winter? It took energy to get the room to that temperature. If we regulate temperatures to minimize the energy we use, we reduce energy demand, which will save money on our energy bills.

Programmable Thermostats

Programmable thermostats can be used in both homes and businesses. The easiest way to save energy through heating and air-conditioning is to turn down the air conditioner in summer months and turn down the heat in winter. If you have a predictable schedule, you can set it up once and then you won't even have to think about it. The most valuable use for your programmable thermostat is reducing usage at peak-demand times. Turn energy use down when you leave your home for work or school during the day, or when you leave your office at the end of the day. Also turn it down when you go to sleep at night. Some utility companies will even provide free programmable thermostats.

There are more sophisticated ways to use programmable thermostats, too. Lights in large buildings are sometimes controlled by motion sensors so that they are turned off when no one is in the vicinity. Motion sensors can also be used with programmable thermostats so that when buildings are empty, thermostats can be adjusted to further reduce energy use. This function can also be combined with lighting controls and security functions.

Shade

In warm climates, providing shade on the sides of the building that get sunlight during the heat of the day can reduce energy demand. This can be achieved by installing awnings on windows, planting shade trees, and keeping curtains and blinds drawn when appropriate. It can be achieved with window treatments, including screens and film on the windows.

Having the roof extend two feet beyond the end of the building will provide a significant amount of shade. There are also creative awning options that take into account the angle of the sun during different seasons to vary the amount of heat radiating on the windows. Free online solar-angle calculators can help calculate the optimum angle. Any awning will make a difference, though.

Shade is beneficial for more than just the buildings. When asphalt and cement get hot, they retain and radiate heat. They are what is called a *thermal mass*, a mass that stores heat and releases it over a period of time. Neighborhoods with lots of trees can be as much as ten degrees cooler in the hot summer months because of the shade.

In cooler climates, you want the opposite, allowing sunlight in to warm the cold rooms. This can be achieved by providing windows on the sunny sides of the structure to allow the maximum amount of sun and corresponding heat to enter the building. You can locate a thermal mass, like a stone fireplace, where it will be exposed to winter sunlight so it stores heat during the day and distributes it slowly throughout the night hours.

When you build a new building, consider the orientation of the house so that you take advantage of the natural direction of the Sunlight and the wind, too. There are many actions you can take to improve the use of shading around your home or business. Explore what will work best for your environment.

Product Heat Generation

We have addressed product efficiency and operation as ways to reduce energy demand. There is another factor related to products. Heat is generated by all kinds of appliances and electrical products, not just heaters! Have you ever stood next to a large-wattage light and felt the heat coming from it? That might be good if there's cool weather, but in warm weather, the heat might increase the need for air-conditioning. This is another reason we need to keep appliances turned off when not in use.

And what about the placement of those appliances? In warm climates, if heat-producing appliances are placed where there's good airflow, it improves the home's overall energy efficiency. In colder climates, place appliances where you can take advantage of their heat generation.

Heating and Air-Conditioning Units

While the factors covered so far contribute to the temperatures in our buildings, the internal temperatures of those buildings are primarily regulated by the heating, ventilation, and air-conditioning (HVAC) units of our homes and businesses.

HVAC units are sized by their tons of capacity. They are also rated according to their seasonal energy-efficiency ratio (SEER).[1] This rating typically ranges from 13 to 23, with the highest numbers used for the most efficient units.

If you incorporate more energy-efficient products and practices into your home and business, those places won't need as much heating or cooling. If you incorporate energy efficiency in the design and construction of the house or building itself, you will further reduce the heating and cooling required. You will find that you can use a smaller HVAC unit, which will cost you less for the unit itself.

As a matter of fact, it is important to get a small enough HVAC system for energy efficiency. HVAC units affect not only the temperature of a space but also the humidity. To remove the moisture from the air, the unit must run long enough for that to occur. If the unit is sized too large for the space, it will blast the air for a short period of time and then shut off before the humidity is removed.

When you combine the reduction in energy used with the increased efficiency when you do use it, you will really make a difference by having an appropriate HVAC system. You will save money on your monthly bills, too! To stay comfortable and guilt free, or at least guilt minimized, an investment in these systems is well worth the money. And they will pay for themselves as energy prices rise.

HEAT TRANSFER

Once heaters and air conditioners have pumped air into your home or business at the right temperature, you want it to stay there! Wherever the boundaries of buildings meet the outside air, there is the potential for heat to transfer in and out. Heat is a form of energy. Losing energy from unwanted heat transfer, whether it is heat getting in or heat getting out, is a waste of energy.

Heat is transferred in three ways: radiation (it travels through the air from something, such as the sun, via electromagnetic waves); conduction (it passes through a solid material, such as a wall); and convection (it is transferred by moving liquid or gas, such as air).

Heat naturally transfers from a hotter region to a cooler region. When you consider how to keep heat from transferring in and out of a building, you might think of it as putting a heat-resistant wrap around the entire structure. The term *thermal barrier* has traditionally been used to mean a barrier to fire. However, it is also applicable to any situation when you want to bar heat from transferring. Having thermal barriers in one part of a wall but not in another will provide an incomplete barrier.

Roofing

The sun beats down on rooftops, so this is a particularly important consideration for heat transfer. The climate of the area in which the building stands makes a difference. Are there more warm months or more cool months? The type of climate affects whether you want reflecting or absorbing roof materials.

Shingles made of composite materials absorb heat from the sun, while metal and tile roofs reflect it. Shingles also serve as thermal mass— they retain heat and then distribute it to the house into the night hours. Therefore, shingle roofs are better in cooler climates. Metal roofs are better in warmer climates where you want to repel or dissipate the heat as quickly as possible.

The sun radiates heat. Therefore, in warm climates, roofing needs some sort of radiant barrier. Radiant barriers reflect heat. Metal and tile roofs themselves provide excellent radiant barriers. For other types of roofing materials that don't reflect as well, a radiant barrier such as a reflective coating can be added.

Of course, if you live in a cool climate, you welcome the heat, so you wouldn't use radiant barriers.

Insulation

These roofs and other barriers will reflect most of the heat, but some will still come through and radiate below the roof. Therefore, additional thermal barriers are used. In addition to transferring via radiation, heat can transfer via conduction. This happens with materials that are conductive, not those that insulate. Insulation basically occurs with materials that do not conduct heat. Metals do conduct heat; Styrofoam doesn't. That means metals aren't good insulators, and Styrofoam is. Wood insulates better than aluminum but not as well as Styrofoam.

Insulation is used in the walls, sometimes in floors, and usually along the ceiling, below the attic. Alternatively, an insulating material can be used above the attic space along the roofline so that the attic temperature will stay consistent with the inside temperature of the house.

There are many forms of insulation. The traditional form is fiberglass batting, which looks like really thick blankets of pink cotton candy. There is a foam insulation that is made from soybeans and looks like a hunk of bread dough! There are other forms of foam insulation as well, some of which are oil based. There are also insulating materials made from recycled newspaper, called *cellulose*, and even a type made of recycled blue jeans!

Insulation is rated to indicate how effective it is at minimizing heat transfer. Perhaps the most common rating for insulation is the R-value, which measures thermal resistance—the material's total resistance to the passage of heat or cold.[1] The higher the R-value, the more effective its insulating properties.

R-values are additive, meaning that if you add another layer of insulation, it will increase the overall R-value. Building codes set minimum R-values for insulation, depending on the climate.

To create our "wrap" effect, roofs, walls, and floors must be considered. They ensure that we completely enclose the envelope of our building. If materials with very good insulating properties are used correctly for part of the surface but are paired with materials that are not insulating or are not installed correctly, it will not have the desired effect. If there are holes or gaps between materials, that will reduce the effectiveness. Two common places that are sometimes overlooked are behind bathtubs and in the corners of buildings. Those pink fiberglass bats also have significant problems with this because gaps easily occur between and around the sections of the insulation.

The R-value of a material alone will not assure us of a good insulating effect. We need to take into account all of these factors.

Windows and Doors

Windows create an interesting challenge because they are like a big hole in our carefully created insulating envelope. Windows are wonderful, though, because they allow us to vary the heat, light, and airflow. They let natural light inside our buildings. Besides reducing our need for artificial light, natural light is good for our health and our productivity.

There are four main considerations with windows: heat radiating from the sun, heat conducting through the windowpane and frame, light, and airflow. First let's consider the heat radiated from the sun.

Windows are rated according to a solar-heat–gain coefficient (SHGC), which indicates how the temperature inside your home or office will be affected by the heat of the sun. Lower SHGC ratings are good when you want to keep the heat out. Higher ratings are better when you want to let heat in.

There is another rating, called the *U-factor*,[1] which measures non-solar-heat flow and includes the glass as well as the frame and spacer materials for windows and glass doors. The lower the U-factor, the more energy efficient the window, door, or skylight. The glass in the windows themselves transfers heat via conduction. No matter what climate you're in, you want to ensure that your window glass doesn't conduct heat because that will let heat escape from the house in cold climates or let the outside heat enter in warm climates. Windows treated with a coating, like low-E (emissivity) or other types of thermally treated windows, will assist in stopping the heat from transferring while still allowing light to pass through. Plus, we can use curtains and blinds as another layer to regulate the amount of light and the resulting heat.

You also need to consider the frames of the windows. For instance, aluminum is a material that conducts heat; it allows heat to pass through. If windows have aluminum frames, they are not providing a good thermal break or barrier. However, if the aluminum surfaces are layered with a material that does not conduct heat, the frames will be energy efficient. This can be achieved with a thermal break internal to the window construction or with a nonconductive decorative layer.

Additional layers provide additional insulation, as we covered in the previous section. Double-paned or even triple-paned windows reduce the amount of heat that transfers through while still allowing the light through.

Consider climate, orientation, and external shading in your decisions about windows. In general, if you want to heat the inside, you want windows on the west and south sides of a building, and if you want to keep the heat out, you want to cover those west and south windows, and have more windows on the east and north sides. Place windows where there is no shade to let the heat in, and place them where there is shade when you want to keep heat out.

Windows give us the option of allowing airflow when we want it, so we need to also consider their placement. We can generate airflow by opening windows on either side of the room, thus cooling a room down. High windows can be used to vent heat. Placing them in stairwells also helps, since heat rises.

Weather Stripping

The third type of heat transfer is convection, or the transfer of heat due to the movement of fluids or air. It's amazing how much heat can transfer through small openings! Heat is lost and gained very quickly. Convection is a useful method of removing heat when you want to, but it isn't good when it is happening where you don't want it. And remember, if your envelope isn't completely closed, you're negating some of the benefits of your insulating materials.

Caulking and weather stripping can ensure that heat doesn't transfer through openings along windows and doors or those meant for wiring. And they're inexpensive and relatively simple to apply and install.

Duct Sealing

Heating and air-conditioning systems use ducts to transport the warmed and cooled air in a controlled form of convection. If these ducts aren't adequately sealed, however, there will be energy loss and unnecessary cost. There are tests to determine if there is leakage in your ducts. Sometimes your local utility company will provide the testing service or will give rebates for it.

Besides the seals of the ducts themselves, the sealing of the space around the HVAC unit also makes a difference. Let's say you have a terrific air-conditioning system, but the air ducts are placed in the hot attic without being well insulated. Your HVAC system is then not as efficient as it could be, which means it will cost you more money to operate than it needs to.

One other option is to use a ductless system such as a "mini-split" system for smaller applications. This is similar to those used in hotel rooms, but the condenser is outside, so it isn't as noisy as hotel units can be.

AIR QUALITY

While you may not want the heat to transfer, you do want air to flow. If you've wrapped your building up really well, you don't want to cause problems by sealing it up in a way that traps moisture, stale air, and/or unhealthy air.

When you cook, you need to have a way for the odors and heat to escape. You need a means for the steam from the bathrooms and laundry room to escape, too. Vents are used to carry the air from these rooms to the outside. Generally, the vents go out the top of the roof. But those aren't the only rooms that need airflow. Remember, humans and pets create CO_2 when we breathe, and create methane gas. Humans also sweat. Airflow and ventilation are important throughout our buildings.

You can use fans to circulate air. You can use windows and doors to provide a nice breeze through your air space. In addition, air needs to be exchanged in a controlled way throughout the building. Controlled air circulation is achieved by the HVAC system. Air is heated or cooled and then transferred throughout a building through the ducts. Over time, the temperature of the air needs to be adjusted again. A return vent draws the air back through the system and the heating or cooling process repeats.

What if the ducts aren't large enough to provide the necessary airflow? If so, it will negatively affect your air quality. Well-designed ventilation is a very important component of any HVAC system.

In large-capacity buildings, there are lots of people exhaling CO_2. CO_2 sensors are used to assess how much outside air is required to maintain good internal air quality. The capacity must meet the maximum occupancy of the building based on 20 cubic feet per minute per person of airflow.

The introduction of outside air is controlled in these larger buildings with heat- and energy-recovery ventilators. Heat-recovery ventilators transfer only heat, while energy-recovery ventilators transfer heat and moisture. Fans bring fresh outdoor air into the building and stale indoor air is exhausted out. The outside air is heated or cooled before it enters the building space.

For houses, air exchange is very simply achieved when doors and windows are opened and closed, but controlled introduction of outside air using the above technologies may also be used. Alternatively, independent air purifiers can be purchased to ensure air quality. These considerations become more important in buildings that are well sealed.

Moisture

Too much moisture can invite mold growth and too little can cause viruses to breed. The optimum level of relative humidity is between 30 percent and 60 percent in most buildings. Vapor barriers may be placed between the insulation and the more humid air to keep moisture out, but vapor barriers can also cause problems if used in climates where they are inappropriate.

Humidity can be monitored and controlled just as a thermostat is used to monitor and control temperature. Humidistats monitor and add moisture to the air. Dehumidistats monitor and take moisture out of the air. Hygrometers are used to monitor the relative humidity without controlling it. Small humidifiers and dehumidifiers can be operated manually for smaller spaces.

Humidity control dampers may be included in the HVAC design so that the damper can be closed on humid days but opened to allow outdoor air to flow on other days.

Attics

Remember when I said that some of the heat from the sun passes through the roof into your attic? And that there is generally some sort of insulation between the inside space and the attic so that the heat doesn't pass into your building? That leaves heat trapped between the roof and the insulation. Attics are generally vented so that that heat can transfer via convection—by the flow of the air.

In humid climates, allowing moist air to flow through the attic may cause more problems than it solves. Therefore, attics may be thoroughly sealed so that moist air doesn't enter the space.

Many businesses and some homes are including their ductwork inside the heated and air-conditioned space instead of the attic areas so that there isn't a significant difference between inside and outside the ducts, minimizing heat transfer. If you don't like the look of exposed ductwork inside your rooms, an alternative is to heat and cool the attic space by insulating at the roofline instead of at the attic floor.

That's a lot to think about! It doesn't have to be overwhelming, though. There are varying levels of energy efficiency and air quality. All of the components of any building must be considered as a whole to ensure that the pieces work together in the optimum way. It's an interesting puzzle, and we just need to ensure that we put the pieces referred to in this section together to create a good end result.

Energy efficiency can save you money and benefit your community and the planet. Better air quality can have benefits on a personal-energy level, too, since it can improve health.

REUSED, RECYCLABLE, RENEWABLE, AND CLEAN MATERIALS

Decisions about construction materials affect not only heat loss and retention but also electricity demand and emissions. I have already addressed the general practice of reducing, reusing, and recycling. These principles apply to construction materials as well. Let's see how choosing materials with a focus on being reused, recyclable, clean, and renewable can make a difference.

Reused

If you're remodeling or tearing down a building, consider which of the old materials can be reused in new construction. Call Habitat for Humanity, an international organization that builds homes in partnership with people in need, or a similar organization, so they can take the materials you won't be using. And if you're building a new building or adding on, consider using materials from a Habitat for Humanity–type store for your project so your purchase can benefit the charity.

Wood flooring, doors, and windows can be reused. Of course, you do want to make sure that they are energy efficient, or you wouldn't want them to be reused! Plumbing fixtures, lighting fixtures, tubs, sinks, cabinets—there is the possibility that these things can be reused in another home.

In addition, warehouse shelving, office furniture, and other industrial and commercial materials should be considered for reuse.

Recyclable

While wood is traditionally used for residential construction, it is not rapidly recyclable. Steel, on the other hand, can be melted down and recycled for new uses. Steel is a conductive material, so you would need to incorporate a thermal break for it to be effective. But if this is done economically, there is no reason that steel couldn't be used on houses instead.

Alternatively, wood offcuts are being used to make finger-jointed studs for housing. What this means is that instead of discarding smaller leftover wood pieces, they are recycled and joined together to make stronger structural pieces for building.

Here's at least one way to reduce the negative aspects of coal-burning for electricity production: The ash from coal power plants is combined with cement to make a better product for building foundations.

Another area in which it is possible to use recycled materials is insulation for minimizing heat transfer (as already mentioned). Examples are cellulose, which is recycled paper, or there are those from recycled denim jeans!

By the time you read this, it's highly likely that there will be many more innovative ideas. Keep researching.

Renewable and Clean

Materials that are rapidly renewable are preferable to those that are not. Bamboo is used for flooring. Bamboo is a grass that can be replenished in five years. In contrast, it takes a century to grow the wood for hardwood floors. There are soy-based foam insulation products that also come from sources that can be replenished quickly.

Materials that do not produce emissions are also preferable—both for the health of those living in the home or working in the business and because of the materials' minimal impact on the environment.

Some building materials are known as volatile organic compounds, or VOCs. Organic implies that they contain carbon (one of the components of carbon dioxide). Volatile organic compounds are chemicals that evaporate easily at room temperature. These can be especially problematic for young children, the elderly, and those who have respiratory ailments such as asthma.

Some of the traditional construction or remodeling products that emit VOCs are paint, carpeting, and vinyl flooring. There are non-VOC and low-VOC paints, and there are alternative flooring options that can be used to reduce these emissions. Varnishes, upholstery fabrics, and some cleaning supplies also may emit VOCs. Eliminating these from the environment will improve air quality and health.

OTHER TECHNOLOGIES AND PRODUCTS

New Products

New, more efficient products are reaching the marketplace every day. Some are improvements on current technologies, such as low-water-flow toilets. Some take existing technology and put it together in new ways, such as electric lawn mowers. Some use newer technology, such as solar thermal water heaters and solar panels that provide electricity directly from the sun.

Existing products are also being modified to be more energy efficient. For instance, there are space heaters designed to help you save on your energy bills in addition to a wide range of energy-saving appliances and products.

If you think about your own environment and your own habits and do a little research to find out what is available, there will continually be new and exciting products coming onto the market to help you save energy and money.

New Technology

There are new technologies that specifically address energy efficiency and the reduction of emissions. There is also a whole new field of science, called *nanotechnology*, which concerns the manufacturing of materials measured in nanometers. (The head of a pin is 1 million nanometers wide.) The potential applications for nanotechnology cross many disciplines, including energy.

Physical characteristics change substantially at such a small scale. Dr. Richard Smalley, co-winner of the 1996 Nobel Prize in Chemistry for the discovery of "fullerenes" (carbon atoms that Buckminster Fuller predicted existed), theorized that carbon nanotubes are the strongest fiber that can ever be created.

Here is an example of how this new field can be applied to energy conservation. Picture a wooden deck with a fence around it and a railing made of many pickets, or thin pieces of wood with spaces between them. Compare that to a solid wall. Imagine painting the two different surfaces. There is a lot more area to paint with the pickets than there is with the solid wood—there is a lot more surface area per volume of space. Now imagine that energy can be stored or captured on a surface. That means a lot more energy can be stored on the picket fence than on the solid one.

Add to this a second property. If dust collects on a surface, it can stop the energy from adhering. However, on a nanoscale it isn't a problem. Dust particles can't adhere because there isn't a large enough smooth surface to adhere to.

Now combine these two properties. If materials are made bumpy rather than smooth, that is, more like the picket fence than the smooth wall, and are also on a nanoscale, more energy can be stored in the same volume of space, and dust won't cause dependability problems. Nanotechnology can provide possibilities that aren't available on bigger scales.

6

THE ECONOMY

THE VERY FIRST DAY of my MBA economics class, the instructor talked about the fundamental assumptions in the field. I remember thinking at the time how flawed some of those assumptions were. In my work as a business-process-improvement consultant, one of the elements we look at is the measuring system itself. It occurred to me at the time that we needed to take a closer look at how we measure the success of our society.

I also think about the economics of my own life. I've had the good fortune to have experienced a wide range of financial situations. It has made me realize that you can be unhappy with or without money, and you can be happy with or without it. A lot of it is in how you look at things.

The way we look at the economy as well as the energy situation is focused on what is wrong with it. That breeds fear and discontent. Fear is the last thing we need. We need to focus on solutions, on what's possible.

I don't think the key to a strong economy is money; it's energy.

KEY ECONOMIC INDICATORS

One of our what-ifs for a solution to the energy crisis and global warming concerns the economy. The economy is the system of human activities related to the production, distribution, exchange, and consumption of goods and services. On a personal scale, it is the standard of living we can afford. What if the energy solution could actually provide a better standard of living for everyone?

In fact, it can. How? First let's look at how the cost of energy affects the key economic indicators.

There are several ways to measure the health of the economy. One of these is called the gross domestic product, or GDP. Other common measures we will look at are inflation and unemployment rates.

Gross Domestic Product

The GDP is considered the most comprehensive measure of an economy's output. It is determined by the value added at each stage of production.

Let's say you have a lemonade stand. Your glasses of lemonade are called the finished goods. Your contribution to the GDP would be your sales minus the cost of raw materials—the lemons and water and sugar —and minus the cost of unfinished goods—cups. If you and the cup manufacturer both counted the cups' value as part of your GDP component, the value of the materials would be double counted. Therefore, the cup manufacturer claims the cup as his output. You claim just the value that you added.

The GDP of the United States is the market value of goods and services produced by labor and property in the United States. It is defined by this equation:

GDP = consumption + government spending + (exports − imports)

We'll look at this in greater detail later, but for now, just understand that GDP measures the value of output.

There are two subcategories that affect this number. GDP can increase because more goods and services are being produced, or it can increase because prices have gone up. Therefore, economists track the change in prices, too, so they can separate the actual consumption from the price inflation.

Inflation Rate

Inflation is a measure of the increase in prices. A closely watched index for inflation rates is the consumer price index, or CPI. The Labor Department has identified four hundred items that buyers typically use; this is called a *market basket*. Each month price surveyors check on the prices of these items in cities across America. These results are then used to compute what the market-basket costs are as compared with what they were in an earlier base period. The percent of change is the inflation rate.

Let's look at how the cost of oil affects this. If the price of oil goes up but the value provided stays the same, that means the price of oil is inflated. As with any product, the price is driven by the cost of producing, distributing, and marketing it. Electricity is needed for production. Petrol gas and diesel are the primary transportation fuels. Therefore, a rise in the cost of oil, and energy prices in general, is likely to inflate the prices of virtually all of the goods you buy, without providing any additional value to you. The result is a rise in the inflation rate.

The quicker we transition to clean, renewable energy that is also affordable, the better off we'll all be.

Unemployment Rate

The unemployment rate is defined as the number of unemployed persons divided by the size of the labor force. The labor force is the number of unemployed persons plus the number of employed persons. A person is considered employed if he or she works at least one hour per week. An unemployed person is anyone who is looking for a job and works less than an hour, or not at all, and is still receiving unemployment benefits. Therefore, the homeless are not included in the number of unemployed.

A weak economy, as determined by GDP, can cause unemployment. Unemployment is also caused in part by a decrease in the demand for goods and services, because fewer workers are needed when fewer goods are being produced.

Inflation can also have an adverse affect on employment, because companies look for ways to compensate for higher production and distribution costs, and one way is to reduce their labor costs. Since labor costs are the largest portion of any company's expenses, first cuts are made there.

If more people are unemployed, spending decreases, of course. The spiral into relative and real poverty gets worse. The unemployed, who are highly likely to be lacking health-care benefits, may suffer more illness and depression, which can have a negative impact on society.

Unemployment is also affected by a mismatch of skills—there may be jobs available, but the available workers, including graduating students, don't have the necessary skills. As jobs move to new technologies, workers may not be qualified to do them.

As new technologies for energy move into the mainstream, there are many opportunities for new jobs. Training for these skills becomes very important. So ramping up new energy technologies and the resulting new jobs will help counteract the potential displacement a changing economy brings.

ECONOMIC ASSUMPTIONS

The study of economics is based on fundamental assumptions. This is necessary to be able to make general observations and to draw conclusions and to compare results year to year. However, if the assumptions are flawed, then the conclusions will be incorrect.

Before we look at how to improve the economic indicators, let's first look at some of our assumptions and our ways of measuring the economy.

Distribution of Scarce Resources

A basic premise of economics is that it is about the distribution of scarce resources. If a resource is scarce and the demand goes up, the price will go up. For instance, there are only so many apples grown each year.

If doctors start prescribing an apple a day for all of their patients, the demand will increase, but the supply in the short term will be fixed and apples will become a scarce resource. The competition for those apples will make them more valuable, and people will be willing to pay a higher price for them. After all, an apple a day is just what the doctor ordered.

As prices get higher, people will look for an alternative to apples that will serve the same purpose. Maybe kumquats would be found to have the same health benefits. Once customers find alternatives, the demand for apples will go back down. If the demand goes down low enough, there will be an excess of apples again. The price will adjust back down since apple sellers will be competing for people to buy their apples instead of someone else's kumquats.

Alternatively, if apples remain in high demand, more suppliers will start to grow them. Eventually, apples will no longer be scarce and apple suppliers will start competing for the sales, so they will adjust their prices downward. Again, the price will adjust so that the supply equals demand.

What if the supply becomes unlimited? Remember, energy is not scarce. It is neither created not destroyed; it just changes form. Therefore, the supply of energy is nor limited; it is just that the desired form of the energy may currently be limited. What would happen if the scarcity of oil and other forms of energy went away? What if we were able to get energy into whatever form we needed whenever we needed it?

There is a precedent for this. Until the 1900s, salt was a scarce and critical commodity. Salt preserved food, which reduced the seasonal availability of food and allowed travel over long distances. People were paid in salt. Wars were fought over salt.

As technological advances were made, salt could be mined more easily, increasing its availability, and other methods of preserving food came along—primarily, refrigeration. Salt was no longer a scarce resource in high demand. Instability went away. Today, salt is easily obtainable and very inexpensive.

Likewise, moving to less-expensive sources of energy could ultimately have a tremendous positive impact on our lives. Since we won't be spending as much on energy, we will have money available to spend on other things.

But if we're spending less on energy, the GDP will look worse. We need to see if what we're measuring is a good reflection of the health of the economy.

Measuring Consumption

The GDP measures the economy based on consumption. In statistical and metrics-oriented circles, there is a saying that we get what we measure. If we measure processing time, it puts a focus on time and promotes the idea of reducing the processing time. If we measure defects in paperwork, it puts attention on the quality and promotes practices to ensure that defects don't occur. Since a higher GDP is considered an indication of a healthier economy, measuring GDP, and thus consumption, prompts us to look for ways to increase consumption. Is consuming the planet's resources really what we want to promote?

If we spend more on medicine, is that better? If we spend more on food, does that mean that eating more is necessarily better? If we spend more on energy, does that mean expensive energy is better than less-expensive energy? Interestingly, war is good for the economy. If we spend more on defense, is that better?

If we use new vehicles that reduce our use of petrol gas, our spending on petrol gas will go down by billions of dollars, reducing petrol gas's contribution to GDP (see figure 6-1). Does that mean we shouldn't do it?

Of course not.

When a child is young, it needs to grow, and it needs to consume to grow. When a business is new, it needs to grow. When a country is not yet developed, it needs to grow. But at some point, growth is not the primary goal. Maintaining healthy energy flow is a better goal.

It is possible to keep the GDP high while still reducing our consumption of physical resources. If spending moves to more service industries, the economy can remain strong without consuming the Earth's resources.

If we're going to measure based on consumption, we at least need to factor in the quality as well as the quantity of what is being consumed. Do we spend our money and energy on things we really enjoy? On

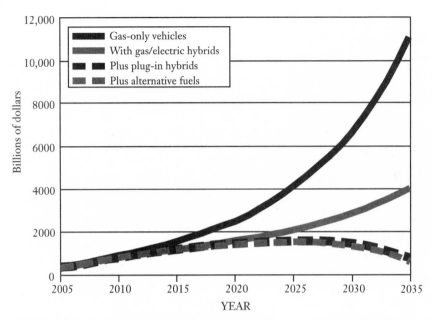

FIGURE 6-1: Total Retail Automobile Fuel Expenditure in the United States

things that improve our quality of life? We need to assess whether measuring our economy primarily on consumption is the right approach.

Centering and Variation

The economic indicators are taken in the aggregate, which means they are taken for our whole economy. Yet what we're really interested in personally is our own individual economics and whether we have the money to pay for the things we want.

If we look at the average temperature for the entire United States, we might say that it's really great. But it doesn't mean much for us as individuals. We don't care about the average temperature of the country. We care about the temperature where we are.

We do need a way to assess what's happening overall, and using an overall measure like the GDP provides us with very useful information. The question is how to also show how such figures vary from person to person. One way is to assess the economy by its variation. You can measure the variation by comparing the highest and lowest value of the measurement. There are other more complicated (and generally more useful) measures as well.

Are the average economic indicators moving up because everyone is moving up a bit? If so, the variability will stay about the same. What if the average indicators are improving because a few people are making significantly higher amounts, while a lot of other people's financial status is getting worse? A measurement of the variability will show this because the variation between the highest and lowest earners will increase.

Just as we need to measure the economy in terms of quality as well as quantity of consumption, we also need to measure variability as well as averages.

The Value of Money

Another basic premise of economics is that the value of a dollar to one person is equivalent to the value for any other person, yet this is obviously not the case. Ten dollars may seem like a fortune to a homeless person or a young child, but to a millionaire, it's not much. Money does not have an absolute, fixed value. A monetary system is merely an agreement on a way to exchange value.

Money provides the storage mechanism for value. Let's say I want one apple for my apple-a-day prescription. I make wagons. The apple grower doesn't want a wagon and I don't want hundreds of apples right now. We need a value holder to store the difference in the value of the two items. Money is a convenient value holder to store the difference in value so we can carry that money to someone else and exchange it for something else. By establishing a value for each item, and a value holder that we agree to use to represent a unit of value, we are able to agree on an acceptable exchange of goods. As long as people agree on the value of that currency, it works.

Since the value is just an agreement between people, it can change, and it has. At one time, you could buy a bushel of apples for a dollar; now to buy one pound for a dollar, you would have to catch a terrific sale. Yet the inherent value of an apple hasn't changed.

The value of a dollar, by contrast, changes constantly, in a variety of ways. It changes in comparison with other currency, for instance. Euros, yen, pesos, and dollars represent agreement on units of exchange within societies. We can exchange a dollar for a Euro or other currency, and just as the exchange of goods is based on a value we assign to each individual item, the value of different currencies is dependent on the value we assign to each of them. Depending on the fluctuating value of different currencies at a given moment, the value of the dollar goes up and down.

All of these examples show the flaw in the assumption that a dollar for one person is the same as for everyone else.

Energy Flow

The value of an item is generally based on the energy it represents: the energy required to create it and deliver it and/or the energy value assigned to it by the recipient. Therefore, money not only stores value, it is a form of energy storage. Money stores the value of the energy of the item. What truly represents our economic health is the flow of energy, whether it's money or other forms of energy. We measure our economy by the flow of money in the purchase of newly produced goods. The Commerce Department does not count the sale of secondhand items in the GDP. These were counted when they were originally purchased. To count them again would constitute double counting. The total expenditure in all transactions is much larger than the GDP.

Yet these secondary transactions are an important indication of the energy flow within our economy. If instead of putting unused furniture in storage you sold or gave it to someone who would really value it, that person's quality of life would be improved by obtaining those secondhand items. By clearing out clutter while potentially bringing in money, your quality of life could be improved, too. What this means is that we increase the overall wealth by providing someone else with something he or she finds valuable.

GDP also does not include financial transactions, such as the sale of stocks and bonds. These involve sales of ownership or debt and do not spring directly from the production of goods or provision of services. Yet they do indicate movement of energy.

Money is a very valuable tool for energy storage and exchange. It is a tool that helps us have a high quality of life by regulating the flow of energy. It is not a goal in and of itself.

COMPONENTS OF GDP

Now let's look more closely at the GDP and how our sources of energy affect the different components of GDP. The GDP measures the energy flow of new production from government spending, consumer and business spending, and exports minus imports. For the purposes of this section, we will assume that consumption is a good basis on which to assess our economy.

Government Spending

Government spending includes funds allocated to education, roads, and health care. It includes pensions for retired and disabled workers and welfare for the impoverished. It covers Social Security, national security, and protection in our neighborhoods. And it includes the interest payments on money borrowed.

If we consider how moving to new forms of energy would affect government spending, we see that the number of government dollars spent on renewable and/or clean-energy products, such as electric vehicles, would positively affect the GDP. A new infrastructure is required to support new transportation solutions, which could include electrical-recharge stations and a hydrogen infrastructure. This would also contribute to the GDP.

The government would need to develop the infrastructure for the energy grid. That includes installing transmission lines from new power-generation stations to substations and reconfiguring existing stations. It also would include replacing outdated metering systems and installing controls and sensors to help implement the smart grid.

The GDP will also be positively affected by government funds spent on energy research and development for subsidies to companies that provide energy products and services that are in the public's interest, and rebates to consumers for buying energy-efficient and emission-free products. Subsidies can take the form of grants, tax breaks, and trade barriers. Organizations get subsidies in the form of grants for research and development. Companies get tax breaks for providing products in the public interest. For example, tax breaks are currently provided for the purchase of electric vehicles and other energy-efficient products.

Government funds used to provide relief on energy costs for the poor and disabled also contribute to the GDP. Programs for providing light bulbs, refrigerators, and other energy-efficient products to replace old inefficient ones will benefit the community and reduce the financial burden on those who can't afford it.

Government-funded educational programs on new energy technologies and energy awareness will also contribute to the GDP.

Consumer Spending

It seems that reducing energy use would negatively impact the GDP. Many of the things we've covered—spending less on energy, reducing waste, improving efficiency—mean consumption would go down. The reality, however, is that the money not spent on energy would probably shift to other goods and services, including new energy-efficient and emissions-friendly ones.

Fear freezes spending. The biggest threat to the economy is fear. When we're frightened by something, we draw in and hold our breath. That isn't a healthy approach to improving our lifestyle. Our human energy needs to flow, and the exchange of money is a good indication in our society that it is doing just that.

The vision of a bright energy future would enable us to breathe more freely, and the GDP would ultimately grow as a result.

Exports and Imports

Exports are products and services we sell to other countries, and imports are products and services we buy from other countries. If we export more than we import, the amount of money coming in is greater than the amount going out, and the difference is added to the GDP.

As a country, we need to look for opportunities to export more than we import. In fact, energy can give us that opportunity. It's obvious, then, that reducing our importation of foreign oil and other forms of energy is good for our economy. If we can fulfill our own energy needs internally, our economy benefits. If we continue to increase our energy production beyond our own needs, we can stop importing and increase our level of exportation to other countries.

Electricity and other energy sources are not the only products we could export. Think of all of the products and appliances that require electricity. If electricity is made available to developing countries, it could increase exports of other electrical products as well. Therefore, developing energy production capabilities is crucial from a global marketplace standpoint.

INVESTMENTS

The GDP formula is sometimes shown with capital improvements or expansion of capacity broken out from consumption. Recall that the formula currently relates to first-time investments only; it does not refer to trading stocks and bonds because that would not reflect production of an output.

Investors can be individuals, organizations, or corporations. U.S. companies had $633 billion in cash and cash equivalents at midyear 2007, with Microsoft leading the way with $34 billion, followed by Exxon Mobil with $32 billion. Just as it is wise for individuals to keep money reserved in savings, it's also wise for businesses. Those reserves can be sources of investment for energy technologies, infrastructure, and energy services and education.

Investors focus on minimizing risk. A factor that has stalled investing in some of the new energy technologies is that tax breaks typically last just five years. Investors may be concerned that it will take longer than five years to realize the returns they desire, and tax breaks won't be extended. While the government can lengthen tax breaks to deal with this concern, this concern is shortsighted.

In new energy-related industries, including some we have covered and those we will cover in the next three chapters, it's not a matter of *if* these products will come to pass; it's a matter of *how quickly* and *who* will make theirs most successful. It's not a matter of *if* the technology will be viable, but *how* to make it so. Whichever companies have good products and are able to grow fastest will have the largest potential for long-range profits, and investors willing to take the biggest early risk and establish their technology as a primary solution will have the potential to reap the biggest reward. There is enough energy demand to go around!

Truly, we are all investors. Anyone can invest his or her energy in the solution. We can invest in stocks that focus on clean and renewable energy, and we can invest our personal energy. As previously noted, money is only one form of energy storage. For the very enterprising, there are opportunities to create new businesses. Besides the major types, such as those supplying solar or wind energy or manufacturing electric vehicles, there are many derivative products. For instance, some electric lawn mowers have power cords. A derivative product could be a device that keeps the cord up, away from the blades. Another might be a type of cord that retracts and extends.

To identify derivative product and service ideas, educate yourself. You might buy an electric lawn mower and see how you can make it easier to use. Attend conferences on green-building and clean, renewable technologies.

Perhaps this is the biggest challenge: with so many different possibilities, how do you choose? The best way is probably to consider what you're really interested in and passionate about. As you've been reading, what areas have interested you the most? As you read the remaining chapters, ask yourself the same questions.

7

ENERGY SUPPLY

WHEN I FIRST MET the six-foot-four-inch Mark Kapner of Austin Energy, he was driving an electricity-propelled Pinto with about eight large batteries in the trunk. He was also one of the most knowledgeable people in the city about renewable energy. In energy-conscious Austin, Texas, that's saying something. Mark was patient and generous with his information, and it was obvious that green was more than just a job for him. Mark provided me with a wide range of information on many aspects of the energy picture and introduced me to other experts in their fields.

In my less-informed state, I believed that the solution to the energy crisis was to merely stop using fossil fuels and start using renewable energy. When I began to understand the numbers, I realized it was a bit more complicated. I also realized how important it is for us to work together, because it isn't a matter of deciding on the one energy type that will solve everything; we need all kinds of energy.

The law of conservation of energy states that energy is neither created nor destroyed; it just changes form. Therefore, it is a matter of having the right form of energy for the purpose intended. It's pretty interesting to understand about all of the different types of energy in our universe.

The information in this section doesn't affect my daily life directly, but it definitely affects me every day because it is about where I get the energy I use every day. If it isn't interesting to you, then skip to the next chapter. But I can tell you that it gave me the framework to see a solution and reduce my own fears about the future. It also reaffirmed to me what an amazing planet we live on. In addition, this chapter and the two following it will show you where you might consider investing your money and/or your human energy, since they are emerging elements of the energy solution.

TYPES OF ENERGY

There are a few forms that usable energy can take in the physical world: mechanical, chemical, and electrical; light and heat; and physical mass. These categories are valid whether we talk about global or personal energy. For instance, think of the forms of energy we use to move or repair our bodies: exercise and other forms of movement, physical therapy (mechanical); food, pharmaceuticals, herbs (chemical); acupuncture and acupressure, which deal with energy meridians in our bodies (electrical). We have energy stored in the mass of our bodies themselves, of course.

Electricity is the main form of energy used to run the machinery of society. It powers our buildings, computer systems, and more. If we want to determine the best sources for electrical energy, we must consider the form of the available energy plus the energy necessary to convert it to electricity. The electric power sector is the largest source of emissions because of the energy required to convert the different forms of energy into electricity.

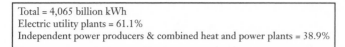

Total = 4,065 billion kWh
Electric utility plants = 61.1%
Independent power producers & combined heat and power plants = 38.9%

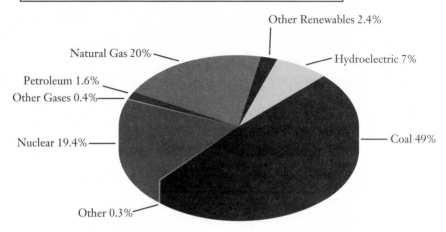

Sources: Energy Information Administration, Form EIA-906, "Power Plant Report,"
and Form EIA-920, "Combined Heat and Power Plant Report."

FIGURE 7-1: U.S. Electric Power Industry Net Generation, 2006

CURRENT ELECTRICITY SOURCES

While petroleum, or oil, is the primary energy used for transportation, it supplies only 1.6 percent of electricity (see figure 7-1). There are other emission-producing sources of electricity, so there is more to reducing emissions and the threat of global warming than reducing our use of oil.

Sixty-nine percent of our electricity needs are supplied by coal and natural gas—greenhouse gas–producing fossil fuels. Therefore, a key question is how quickly we can ramp up non–fossil fuel sources of electricity.

Let's look at an example to understand the challenge. Say you're in the baking business. Currently you bake ten cakes a day. You get a call asking for twenty cakes a day, starting immediately. Can you do it? At those small volumes it might not be too difficult to double your production, because you can just work more hours. But what if they ask you to increase production to two hundred a day? Or two thousand? It would take time to get the ingredients and new ovens.

Let's say the company that is placing the order buys two hundred thousand cakes a day to sell in its stores. Even if you produce two thousand a day, which would be a lot for you, it will be a very small portion of what the company buys.

That's the position alternative-energy companies are in. To make a dent in the current energy demand, they must be able to get the materials and the equipment and the trained staff to produce those higher levels. And that takes time.

It also takes effort on the part of the utility companies or some other organizing body to work with many small suppliers rather than a few large ones.

While there are many ways to make the transition, one thing is clear: There needs to be a portfolio of energy. If multiple types of non–fossil fuel sources ramp up simultaneously, we will transition as quickly as possible to a healthy, sustainable portfolio for our planet.

ELECTRICITY REQUIREMENTS
FOR ELECTRIC VEHICLES

If we move to plug-in vehicles, we will be using less petrol gas, but now cars will need electricity as the source for their energy. If fossil fuels are used to supply that electricity, we may have addressed the problem of oil dependence, but we still will not have addressed the problem of emissions and global warming.

This highlights a key characteristic of the energy situation. It is systemic, meaning there are many systems that affect each other, and all must be addressed to determine how the changes in one area affect another.

In one scenario described earlier, cars will be plugged in at night, increasing the demand for electricity. Figure 7-2 shows the additional electrical power needed if all vehicles are plug-in hybrids: approximately 1,700 terawatt hours (tWh); a terawatt is 1 trillion watts. To put this in perspective, current U.S. electricity consumption is between 4,000 and 4,500 tWh, or more than 4,000 billion kWh.

Does it make sense to increase the demand for electricity then? Yes, but only if we meet the demand using a balanced portfolio of energy.

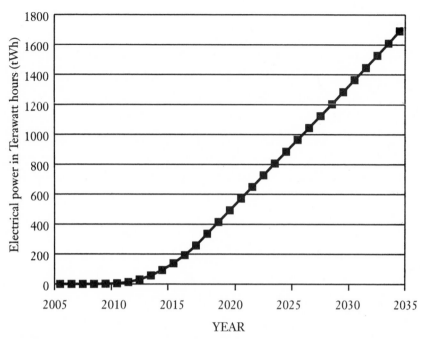

FIGURE 7-2: Electrical Power Needed for Plug-in Automobiles in the United States

CLEAN AND RENEWABLE ENERGY SOURCES

Renewable energy sources do not deplete natural resources. They can be replaced by natural cycles or through good management, and they are Component #4 of our solution.

The Texas legislature has adopted the following definition for energy, composed by the Texas Renewable Energy Industries Association: "Any energy resource that is naturally regenerated over a short time scale and derived directly from the sun (such as thermal, photochemical, and photoelectric), indirectly from the sun (such as wind, hydropower, and photosynthetic energy stored in biomass), or from other natural movements and mechanisms of the environment (such as geothermal and tidal energy). Renewable energy does not include energy resources derived from fossil fuels, waste products from fossil sources, or waste products from inorganic sources."

Green energy is a term generally used for clean, renewable energy, but there is some disagreement about which sources can be called green.

Some energy sources are renewable but are not clean. We will use the terms *clean* and *renewable* to make the distinctions between these two characteristics, where clean means it does not emit greenhouse gases when used and renewable means that it can be replaced in a matter of years rather than decades or centuries.

Just as different climates are better for growing different crops, different parts of the country are better for producing different types of renewable energy. Information about renewable-energy sources throughout the country can be found at www.seco.cpa.state.tx.us/re_maps.htm. As an example, the map on solar-energy resources is shown in figure 7-3.

In this section I primarily addressed energy sources being used to generate electricity for power stations. (Some smaller-scale applications are also covered.) Now let's address some specific characteristics of the various forms of renewable energy.

Wind—On- and Offshore

Wind energy is derived from large turbines similar to the windmills of old. The wind is converted to energy when the turning of the blades causes a driveshaft to rotate, which in turn causes a generator to produce electricity.

Wind energy is currently the cheapest clean and renewable energy. It is also relatively quick to implement, since it does not need a huge processing plant to convert the energy into a usable form. Wind does not require any water to capture and convert it, while other forms of clean, renewable energy do. This can be a benefit, since water is such an important resource for our personal needs and there is a finite amount of it on our planet.

Wind turbines can be located on land or offshore. They can be small enough to serve one household, or they can be part of a wind farm, supplying electricity to a utility company.

There are concerns that the rotating blades of the wind turbines kill birds. Studies are required to ensure that the turbines are not erected along the migration paths of birds.

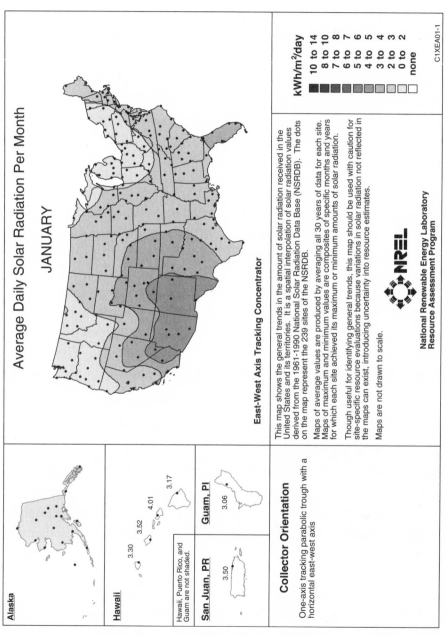

Average Daily Solar Radiation Per Month

JANUARY

East-West Axis Tracking Concentrator

This map shows the general trends in the amount of solar radiation received in the United States and its territories. It is a spatial interpolation of solar radiation values derived from the 1961-1990 National Solar Radiation Data Base (NSRDB). The dots on the map represent the 239 sites of the NSRDB.

Maps of average values are produced by averaging all 30 years of data for each site. Maps of maximum and minimum values are composites of specific months and years for which each site achieved its maximum or minimum amounts of solar radiation.

Though useful for identifying general trends, this map should be used with caution for site-specific resource evaluations because variations in solar radiation not reflected in the maps can exist, introducing uncertainty into resource estimates.

Maps are not drawn to scale.

**National Renewable Energy Laboratory
Resource Assessment Program**

kWh/m²/day

■	10 to 14
■	8 to 10
■	7 to 8
■	6 to 7
■	5 to 6
■	4 to 5
■	3 to 4
□	2 to 3
□	0 to 2
□	none

C1XEA01-1

Alaska

3.30   3.52   4.01   3.17

Hawaii

Hawaii, Puerto Rico, and Guam are not shaded.

San Juan, PR   **Guam, PI**

3.50   3.06

Collector Orientation

One-axis tracking parabolic trough with a horizontal east-west axis

FIGURE 7-3: Average Daily Solar Radiation per Month

Wind is not as reliable as other forms of energy because wind strength varies. This problem is compounded by the supply-demand mismatch: The wind often blows more at night, when demand is low. This means the energy will either be lost or it requires a storage mechanism. There is great synergy between wind energy and electric cars, though, because electric cars can be recharged during the evening hours.

Tidal and Hydropower

Hydropower is created when moving water is used to generate electricity, either on a large scale, such as through dams and waterfalls, or on a much smaller scale. Moving water turns turbines like moving air turns windmills, which can then be used to turn a motor and generate power.

The shifting of the tides is a form of hydropower used to generate electricity. If you've ever been in the ocean, you know that the force of the water can knock you over. That mechanical force can be converted into electrical power. Tidal is a good source of energy in areas that have the right landscape but obviously is not available everywhere. These energy forms are available on demand because tides happen every day, and the water used to create hydropower is always moving. Therefore, these forms of energy are predictable and reliable.

Geothermal

Geothermal energy uses the Earth's heat to generate electricity. While the highest temperatures on Earth are near volcanoes, geothermal energy is present everywhere beneath the Earth's surface.

Geothermal energy can be extracted from naturally occurring steam and hot water. Power plants that use geothermal energy are located near springs or reservoirs that are near the Earth's surface. They can also dig into deeper layers to access the energy.

Geothermal energy is also useful because the Earth's heat is always available and accessible. While temperatures aboveground may change a lot, the temperature below the surface stays almost constant. For most areas, this means that the ground is usually warmer than the air in winter and cooler than the air in summer.

Geothermal heat pumps use the Earth's constant temperatures to heat and cool buildings. They transfer heat from the ground (or water) into buildings in winter and reverse the process in the summer.

Solar Photovoltaic

Solar-photovoltaic devices convert the light of the sun into electricity. Solar cells are arranged on panels. The larger the surface area, the greater the amount of electricity that will be generated. Panels can be used on a large scale, such as on generating plants, or on a small scale, such as on homes and office buildings, on vehicles, or on small devices (flashlights, for example).

The solar cells generate direct current, or DC, which can charge batteries or turn motors. To create electricity for buildings, the energy must be converted to alternating current, or AC.

One challenge with solar energy is that the intensity of sunshine varies, so it is not dependable. As with wind, there must be a way to store the energy to account for the fact that the supply and demand aren't always in sync.

While it is most effective to collect solar energy in sunny climates, solar energy is viable wherever the sun shines. Solar energy is most abundant during the heat of the day, which coincides with the demand.

Solar energy is a relatively expensive source of electricity at present. Unlike with wind energy, photovoltaic devices require a complex production process, which requires a processing plant. Therefore, the initial costs are higher. The cost per unit of energy goes down as the volume produced goes up, because those costs will be divided among more solar cells.

The industry currently has issues with cost and supply of raw materials. Growth of the solar-panel industry is limited by the availability of silicon, which is also needed for computers. However, silicon is the second most abundant substance on Earth, after oxygen. Certain types of sand are a source. Again, it's a matter of changing the substance to the form that is needed when it is needed.

As with all new technologies, the design and costs will improve over time. In addition, new technologies are being developed to achieve the same result with fewer materials, such as a technology called *thin-film silicon*, and to use different types of materials to capture the energy.

One intriguing benefit of solar energy is that the sun is capable of supplying all of our energy needs.

Solar Thermal

While solar photovoltaics convert sunlight to electricity, solar-thermal units convert solar energy to heat. With solar thermal, the sun beats down on tubes containing liquid, heating the liquid. The liquid stores the heat, which can be cheaper and more efficient than current methods of storing electricity.

Solar thermal units can be used for power plants and can also be useful for heating buildings and for heating water.

Biomass

Biomass is a source of chemical energy and is produced from plant material, vegetation, or agricultural waste. The simplest example of getting energy from biomass is when we burn wood or dung to get heat. Biomass can also be converted into biogas for vehicles, as discussed in chapter 2 in the section on alternative fuels.

In addition, biomass can be used to generate electricity. Biomass emits methane gas as it decomposes. Methane is odorless, but it is emitted as part of that smelly gas (sulfide gas causes the smell) that makes garbage and human and animal waste so offensive. Methane gas can be burned to produce electricity. Landfills are sources of methane gas, as are biogases derived from biomass. Methane is also the main component of natural gas, so it is an obvious replacement for natural gas, which we will address shortly.

One advantage of biomass as an energy source is that when it is composed of waste products, it provides a use for that waste. If biomass is left as waste, it breaks down and produces methane anyway. We may as well get the benefits of the process.

There is a limit to the amount of available waste, so scientists are researching crops and algae specifically for creating biomass. If we choose to grow crops to create biomass, one benefit will be that crops consume CO_2 through the process of photosynthesis, as we covered earlier. Therefore, it would compensate for some of the increased emissions. But we must decide whether to use land to grow crops for biomass or solely for food.

Biomass is renewable—we can make more of it very quickly—and it has a benefit over wind and solar energy in that it is also easily storable. Biomass is not a clean source of energy, because it produces emissions. It leaves behind CO_2 and methane—greenhouse gases—when it is burned, either in its original form, such as wood burning, or as a fuel, such as ethanol. Therefore, even if the energy used to produce it were entirely clean, the biomass itself would still produce emissions.

Hydrogen

Hydrogen has the unique capability of being clean, renewable, and storable in the form of water, as hydrogen gas, or as liquid hydrogen. As I've already addressed in chapter 2, hydrogen is transportable, so it can be used for vehicles. It can also be used as a source of electricity. Since it is storable, it can be available upon demand, unlike wind or solar energy, and it is completely clean, unlike biomass.

To be a usable energy source, hydrogen must be separated from other elements that it is attached to. For instance, water is H_2O, which is a combination of two hydrogen atoms and one oxygen atom. Hydrogen can be separated from the oxygen, used to power a motor, and then converted back to water.

It takes energy to separate the hydrogen from the oxygen. It also takes energy to convert it into a liquid form if needed. Therefore, hydrogen cannot fill all of our needs for electricity. It must be produced using energy that is completely clean and renewable, such as wind, solar, tidal, hydro, or geothermal, for it to be considered a completely clean and renewable source of energy.

NONRENEWABLE SOURCES

Since it will take time to ramp up renewables from their current 2 percent share of our energy market, we will need to use the energy forms that are most prevalent today for a period of time. Currently, coal supplies almost half of our electricity, with natural gas and petroleum (oil-based) products providing an additional 21.7 percent.[1] These three fossil fuel sources, combined with nuclear power, provide more than 90 percent of our electrical needs, and all are nonrenewable. Of course, in a few million years or so, they will have renewed, but not in our lifetime.

All of these energy sources raise concerns about toxicity. All but nuclear power contribute to global warming. Let's look at specific considerations for these nonrenewables and how we can minimize the drawbacks while we transition to clean renewables.

Oil

Oil is used for a very small portion of our electricity. As I pointed out in chapter 2, its major uses are for petrol gas and other transportation fuels. We also use it to create plastics, asphalt, and more.

Oil is changed into these usable forms through a process of distillation. To understand how that works, think of it as a series of sieves that has lots of rocks, pebbles, and sand poured into it. The bigger chunks are trapped in one layer, while the rest go through. The next biggest chunks are trapped in the next layer, and so on, until the smallest particles pass through to the last layer.

The various components of crude oil have different sizes, weights, and boiling temperatures. Instead of separating the components by size, using a sieve, they are separated by their boiling temperatures, using a process called *fractional distillation*.

While the process is much different, the end result is similar: Different grades of oil are divided up so that they can be used for the purpose that best suits their properties. To truly get rid of our dependence on oil, we would need to address these other oil needs in addition to fuel for our vehicles and electricity demands.

We've already addressed the fact that oil is a fossil fuel and, therefore, emits gases, which produce a greenhouse effect, which appears to warm the globe, which affects weather patterns. We are dependent on other countries for our supply of oil, and since our relationship with some of these countries is strained, there is cause for concern about the future availability of oil. Just in case that isn't enough cause for worry, this nonrenewable source will eventually run out.

Coal

Coal is used to supply nearly half of our electricity needs. It is burned and used to heat water to produce steam in large processing plants. The steam is then used to turn large turbines. This is similar to the process used with the other fossil fuels from which electricity is derived, but coal doesn't need to go through extensive refining before it's burned.

Coal is currently the cheapest source of electricity, but it is also a fossil fuel and is nonrenewable. It is still considered abundant, so in the short term, we are not as concerned about running out of it as we are with oil.

Coal not only produces carbon dioxide emissions, but also produces two additional greenhouse gases: sulfur dioxide and methane.

There are efforts taking place to reduce the toxicities associated with burning coal. The coal is washed to remove noncombustible impurities. This process is called *sequestration*.

We are likely to still need coal for a while until the clean, renewable forms of energy ramp up their capacity. Therefore, these efforts to make coal more viable are imperative. The effort to convert these toxicities to something useful is imperative, too.

Natural Gas

In addition to being a source of electricity, natural gas is used directly for heating homes and fueling gas stoves. It is combustible, so when it is burned it provides energy. Natural gas in a gaseous form is not easily transportable, and so it is sometimes converted to a liquefied form.

Liquefied natural gas takes up about one-six-hundredth the volume of its gaseous form, so much smaller tanks can be used to transport it.

Natural gas also goes through a distillation, or refining, process to separate its components. It is formed primarily of methane, or CH_4, which means it is composed of one carbon atom for every four hydrogen atoms. Propane and butane, used for gas grills, gaslights, and other purposes, are fuel sources that also originate from natural gas.

As with oil and coal, natural gas is created over millions of years from the fossilized remains of plants and animals. It is usually found in oil fields, coal beds, and natural-gas fields. It is also nonrenewable and produces emissions. However, it produces far less emissions than oil and coal.

Natural gas is relatively efficient, as compared with electricity, which means it provides more energy output per unit of energy input. When the entire cycle of producing, processing, transporting, and using energy is considered, natural gas is delivered to the consumer with a total energy efficiency of about 90 percent, compared with about 27 percent for electricity.

Our sources for natural gas are predominantly from outside the United States. Only 3 percent of the natural gas reserves on our planet are within the United States.[2]

If we are going to reduce emissions of greenhouse gases, we need to reduce the use of natural gas. Since it is nonrenewable, we also need to wean ourselves from our dependence on it. The price of natural gas has also been rising, and as this nonrenewable source becomes more scarce, prices will continue to elevate.

Nuclear

Nuclear energy is produced from uranium, a dense metal. It is not burned to produce electricity, like the other nonrenewable sources. Instead, heat is generated in a nuclear reactor by a process called *fission*. The heat is then used to make steam to turn turbines to produce electricity.

Fission is a process in which uranium atoms are split roughly in half, which releases energy in the form of heat. When this happens over and over again, many millions of times, a very large amount of heat is produced from a relatively small amount of uranium.

Uranium is very plentiful in the Earth's rocks, and much less of it is needed to produce electricity than with the other nonrenewable sources we have covered. While its supply is limited, the end of that supply is not imminent.

Nuclear energy does not come from fossil fuel and doesn't have CO_2 emissions, so it is not a factor in global warming. However, it produces radiation and nuclear waste, which is highly toxic and dangerous. Therefore, it is considered a good alternative to fossil fuels but has its own risks.

As with coal, efforts must be made to reduce the toxicities associated with nuclear power. While nuclear energy may be considered the lesser of the evils right now, our use of it still needs to be reduced.

BALANCED PORTFOLIO OF ENERGY

It is probably evident that we need a portfolio of energy. Renewables need to be ramped up as quickly as possible. With nonrenewables, we need to minimize the toxicities associated with processing them. And with each energy source, we need to improve its efficiency.

When we consider any potential new energy source, we must assess costs and the amount of time required to develop that source. Costs are either fixed (those that do not vary due to number of units sold or produced, such as those for equipment, office space, manufacturing facilities, etc.) or variable (those that change based on number of units sold or produced—for example, the cost of raw materials).

The fixed costs of a wind turbine are very small in comparison with those of a coal plant, since a wind turbine–manufacturing facility is much less complex and less expensive than a coal plant is. There are no variable costs, since wind is free. And a new wind turbine can be produced in approximately one and a half months. There are other practical considerations, such as how to eliminate the challenges of the variability of the wind supply. I will address that issue in the next chapter.

A combination of financial costs plus the social costs—in this case the environmental costs—is a more accurate reflection of the true cost of a potential option.

Figure 7-4 groups different types of energy according to their general characteristics.

The energy sources that are rapidly renewable are in white letters. As you move from the outside to the inside of the circle, you get to the usable forms of energy, or energy carriers, in the center. The outermost ring identifies energy in its raw, natural state, commonly called the feedstock. For the energy sources left of the vertical line, you'll notice that they all go through a conversion process to change them into solid, liquid, or gas form. The benefit of this is that they can then be stored and used whenever and wherever they are needed. These are the refined products.

The energy sources right of the horizontal line are not naturally storable. The good side of this is that it takes less energy to get them

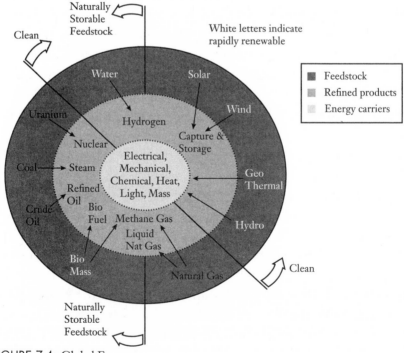

FIGURE 7-4: Global Energy

into a usable form because they don't need to go through an interim stage. The bad side is that they are not storable and thus not as easy to transport. I'll address energy storage next.

You may notice that hydrogen is the only source that is completely clean, renewable, and storable. Yet it isn't the all-in-one answer either, as discussed earlier.

When we consider the rising energy demands in developing countries, it becomes clear that it isn't a matter of deciding which type of energy to use; it's a matter of using a balance of energy sources and determining what is best in each situation, with minimal environmental impact.

RAMP-UP OF RENEWABLES

Let's look at the ramp-up of renewables to see what might be possible. The demand for electricity in the United States grew by annual rates of 4.2 percent, 2.6 percent, and 2.3 percent in the 1970s, 1980s, and 1990s, respectively, but it is projected to slow to an annual 1.1 percent rate according to the Energy Information Administration's Annual Energy Outlook, 2008. This has been added to the projected ramp-up of electricity demand due to plug-in vehicles for the total projected demand shown in figure 7-5.

Currently renewables supply just over 2 percent of the electricity we need. In figure 7-5, the growth rate for renewables is projected to be 35 percent in the early years. As we covered with the cake-baking example, a higher growth rate can't be sustained as easily when the base amount gets higher, so a deceleration rate of 10 percent has been included until it eventually levels off to a 15 percent annual growth rate. With these rates applied, renewables can address 20 percent of the electricity needs, including those for plug-in vehicles, by 2020.

What would it take for the ramp-up to occur faster? Companies increasing their production more quickly and/or more companies getting into production for renewables. If each of the fifty states focuses on replacing its own energy demands, there is enough opportunity for boatloads of investors to succeed.

The motivation for companies to increase the ramp-up of renewables comes from increased demand, either mandated by legislation or

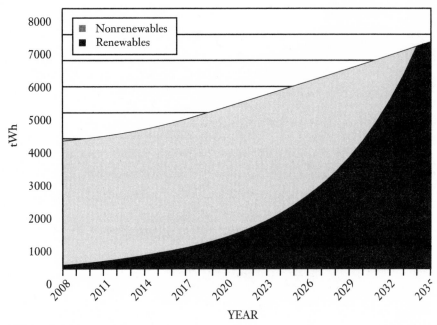

FIGURE 7-5: Renewables

insisted upon by the utility companies and/or consumers demanding it. In the next chapter, we will address another component required to make this ramp-up viable.

The phase-out of nonrenewables needs to focus on oil and natural gas first, for the reason described in chapter 2 and because, as previously stated, only 3 percent of our natural gas is supplied from within the United States. The phase-out of coal and nuclear energy would then follow as necessary.

8

ENERGY STORAGE

I MET CARLOS COE of Xtreme Power, an energy-storage firm, at an event hosted by the Clean Energy Incubator in Austin, Texas. A very sincere, almost shy man, he genuinely believes in the importance of his product; that was easy to tell. Xtreme Power's devices can serve large-scale systems or portable systems you can transport on the back of a truck. The company had attempted to work with electric car batteries but realized its technology worked better elsewhere.

Shortly after that, I met two of the three public utility commissioners of Texas at a legislative session. Barry Smitherman and Paul Hudson sat next to me right after I spoke at the Business and Commerce Committee meeting during the Texas legislative session. I had spoken about a renewable-energy bill. When I realized who they were, I leaned over to Barry and said, "We need more renewable energy." He whispered back, "We need storage." I realized then why Carlos was so enthusiastic about his product—I hadn't previously recognized how crucial storage was to the move to renewable energy sources.

AMOUNT OF STORAGE

Energy supply must match the demand when it is needed, where it is needed, and in the form it is needed. A similar situation occurs with our bodies' need for air. We need a certain amount of oxygen in a week, but the supply has to be available at the right time, too. Inhaling a whole week's worth of oxygen in one hour would not work. The air must be available consistently over the entire week. There may also be periods when we need a higher quantity of air. For instance, if we're going to run in a race or do any other strenuous activity, our intake requirement would increase for a short period of time.

A similar situation exists with energy. Unless we can match the energy supply with the demand at each moment in time, there needs to be a means to store it for later use. Therefore, energy storage is Component #5 of our solution.

Storage can be employed on both the supply and the demand side. For instance, energy suppliers can store energy so that they can provide it to their customers on a more consistent basis. Customers can store it so that they can keep reserves on hand to use whenever necessary. There are also potential uses for storage at distribution points along the way.

Peak Demand

When utility companies provide energy, they must ensure that they can supply it through all levels of demand. More energy is needed during the day, when people are up and about and when businesses are in full operation. More is needed during the hottest or coldest hours of the day, depending on the season. The times when the most energy is needed are called *peak demand*.

Systems that supply energy have a certain capacity. They can provide a certain amount of energy per time period. This is similar to our system of communicating—we have a certain capacity for how many words we can speak in a time period. You can vary the speed at which you talk, but there is a limit to how fast you can talk. If you need to communicate something in a particular time frame, you may have to find a method that will increase your capacity to communicate or spread it over more time or find some other means to meet your objectives.

The same is true for energy. It is critical that the method for supplying energy can handle the peak demand or else there will be brownouts or blackouts, which are terms used for shortages of energy that cause lights and other electrical systems to stop working.

Energy companies use various methods to ensure that they have enough capacity to meet the needs of their customers. As we saw earlier, there are many different sources of energy, and the costs vary. Generally, the least expensive source is used first. When demand exceeds the capacity for that source, another source must be used. The source used to meet peak demand is usually the most expensive one.

Variability

In addition to the magnitude, or overall amount, of demand, there is another key component of demand that drives the amount of storage required: variability. The inconsistency of something is described as its variability. Although you can reduce variability, it always exists. The amount of salt on a potato chip varies from one chip to another; the amount of ink in a pen varies. Our ability to detect these variations reflects the quality of our measuring systems, not the variations' existence.

The demand for energy is variable—it changes from month to month, day to day, and minute to minute. As we've already seen, there is also variability in the supply of some energy sources, such as wind and sunlight. These factors affect the amount of energy storage required.

When we're trying to understand overall energy demand, it's useful to know the average demand per household, per region, per minute. On the supply side, it's useful to know the average watts supplied by a wind turbine. But in metrics and statistics circles, we know that averages are never the whole story.

Let's say the average temperature in a building is 70 degrees. That sounds OK, doesn't it? But if instead the temperature fluctuated throughout the day anywhere from 60 to 85 degrees, would that be acceptable? Probably not. That's a variation of 15 degrees. If that were the environment in which you were required to work, you would have to be prepared to function at both the high and the low ends and everywhere in between. Alternatively, you could see if there were some way the temperature could be stabilized.

In terms of energy, the supply must be able to handle not only the peak demand but also the variability. Another way to think of this is in terms of savings. Savings is essentially the amount of monetary "energy" we keep stored and readily available. The variation of our income and the variation of our expenditures affect the amount of savings we should keep on hand so that we don't run out of money and/ or take on unhealthy debt. In the same way, the variation of incoming energy and the variation of outgoing energy affect the amount of energy that should be stored and readily available so that we don't run out of energy.

In addition to the causes of variability in energy that we've already addressed, another one is weather, and the utility companies must deal with this as well. We cannot control weather, but if we can at least reduce the manageable components of variability, it provides great value to the overall energy picture.

Maintaining a Buffer

A way to deal with variability and ensure that peak demand can be met is to establish a buffer. This means that we always have more than enough energy available. Even if utility companies are set up to meet peak demand, there can be interruptions in the normal supply that require a reserve stock, or buffer, of energy. That's why storage is crucial.

Variability affects how much buffer we need. If we have too little, we may run out. If we have too much, that's not good either because it requires more equipment and space for storage, which means added costs and wasted energy—the energy not being used and the energy required to store energy.

This is like having so much stuff in your house, you have to pay for external storage. In business, you want to keep enough stock to meet customer demand, so you have a buffer. But if you have too much buffer, it costs extra money for inventory expenses. It's also similar to having too much energy stored in your body in the form of fat, which means you use up energy carrying it around. It wastes energy. Losing weight will mean you need less energy to carry excess weight around and save money on food, too!

The buffer needs to match the variability. For instance, if your trip to work or school in the morning takes ten minutes and is along a pathway that is relatively consistent, you can leave at about the same time every day and expect to make it on time. You might give yourself a three-minute buffer to allow for variability. But if your commute takes forty-five minutes and there is a possibility of traffic jams, you probably need to have a bigger buffer. You might allow a fifteen- or twenty-minute buffer, leaving more than an hour before your start time, and then enjoying a few minutes reading the paper or socializing when you arrive early.

As we move to the variable sources of wind and solar energy for our supply, energy storage becomes more crucial.

Leveling the Demand

The more consistent the demand, the easier it is to meet that demand. Consumers can help by doing whatever they can to reduce variability.

Here are some examples of actions we can take. Washers, dryers, and dishwashers can be run during the evening hours instead of during the day, when the demand for electricity is highest. Businesses and individuals alike can, whenever possible, restrict the use of heat-generating equipment to the evening hours during summer and to the daytime hours during winter. We can help reduce petrol gas consumption by staggering work hours so traffic congestion is reduced. There are many ways to smooth out demand of all sorts. See what other ideas you can come up with.

There are significant financial benefits to smoothing out your energy demand, because the per-unit cost of energy is higher during the high-usage hours. And remember, we can make it a game rather than a burden.

Recovery Time and Impact

Another factor affecting energy storage is how quickly you will be able to get more supplies if there is a shortage or outage. This is related to how urgent a consistent supply is. If you have a business in which your product is refrigerated and will spoil if electricity goes out for an extended period, then energy storage is more critical for your business than it might be for some other types. If you run a billion-dollar manufacturing line, the cost of one day of outage could be very high. If you run equipment that keeps people alive, such an outage could be catastrophic.

If you need to be sure you always have enough energy supply, you may choose to have storage on the demand side. In other words, rather than rely on a steady supply from a utility company, you can maintain your own buffer.

This is similar to keeping a small amount of healthy snacks with you so you don't have to find a food source if you get hungry. You wouldn't have to buy expensive and unhealthy foods when you're caught with low energy. It's also like keeping cash reserves so you don't have to borrow money at a high interest rate.

For our global energy solution, it means using energy storage. Besides ensuring that our energy source is not interrupted, this allows us to time our usage so that we don't have to seek more energy at peak-demand times, when costs are so much higher.

These examples of variability show why energy storage is an important component of an energy solution. Assessing how much storage is needed is an art and a science.

TYPES OF STORAGE

The three most prevalent types of transportable energy storage today are batteries, fuel cells, and ultracapacitors (with compressed air as an additional option). Stationary energy storage can also be achieved by pumped hydro and compressed air energy storage (CAES).

Decision Criteria

When we consider energy-storage systems, we must consider how quickly they charge, how quickly they discharge, and how much energy the system can store. We also must consider the maximum amount of power that can be supplied at any given point in time, called the *peak discharge*, and the continuous discharge, which means how much a storage system can provide consistently over a period of time. And we must consider how long it takes to recharge.

We also must be concerned about energy density: how much space the buffer takes up as compared with how much power it will provide. Energy density affects both the size and the resulting weight of the storage containers. Energy density is especially important for energy sources that must be transported. Generally, matter in its solid rather than liquid form provides more energy per unit of space, with gas having the lowest energy density of the three. Natural gas is converted into liquid natural gas because it can be stored in smaller containers and/or transported much more efficiently than if it remained in a gaseous form.

Batteries

Virtually everyone has used small AA, C, and D batteries in children's toys and flashlights and tinier ones in watches, cell phones, and other gadgets. When these batteries are used up, you throw them away . . . unless you buy rechargeables and a charging unit.

Portable computers have batteries that recharge when you plug a computer into an electrical outlet. Car batteries, which are used to start up the engine, recharge when the car runs. Separate, larger batteries are used to run electric vehicles. Batteries for supplying electricity to buildings are obviously even bigger. Yet the principles are the same.

Batteries store electrical energy chemically. Positive charges are kept at one end of the battery and negative charges are at the other end. When the two terminals of the battery are connected to something, it provides a pathway for the electrons to move. Electricity is the flow of electrons through a conductive path like a wire. The electrodes within a battery react and change as a battery is charged or discharged in a reversible chemical reaction.

There are many materials used to make batteries. The commonly used disposable kinds are lead/acid. These are toxic, which is why you should dispose of them properly. Nickel-cadmium rechargeable batteries were the original technology used for laptop computers. Many other technologies for batteries exist, such as nickel hydride and lithium ion. Research and development seeking batteries with better storage properties and lower toxicity is ongoing.

Fuel Cells

While a battery makes use of an electrochemical reaction, fuel cells use liquid fuel to produce electrical energy. The electrons and protons of the reactant fuel are separated, so that the flow of electrons can create electricity. Fuel cells consume the reactant, which must be replenished. There are many kinds of fuel cells. The one we have addressed so far is the hydrogen fuel cell, which uses hydrogen as fuel and oxygen as oxidant.

Because fuel cells have no moving parts and do not involve combustion, in ideal conditions they can achieve up to 99.9999 percent reliability. This equates to less than one minute of downtime in a six-year period. Therefore, fuel cells are especially good for critical applications, for use in remote locations, for aircraft, and for large transportation vehicles. Some are restricted by high or low temperatures.

As I pointed out in chapter 2, hydrogen is storable and has the potential to be fully clean and renewable when coupled with clean and renewable sources of electricity to create it. This is also why hydrogen is part of the long-term solution—the fuel cells can provide fully clean, storable, and renewable energy.

However, energy is required to separate the hydrogen, so hydrogen fuel cells may not be the most efficient storage method for all applications.

Ultracapacitors

Ultracapacitors store charges electrostatically, which means the energy they provide comes from static electricity. When certain materials are brought together and then separated, an accumulation of electric charge can occur, which leaves one material positively charged while the other becomes negatively charged.

An example of static electricity is when you get a mild shock from touching a grounded object after walking on carpet. Excess electrical charge accumulates in your body from frictional charging between your shoes and the carpet. The resulting charge buildup within your body can generate a strong electrical discharge.

Ultracapacitors can withstand hundreds of thousands of charge/discharge cycles without degrading and can provide quick bursts of energy. But they are currently more expensive per unit of energy provided than batteries are.

Other Power Cells

There are ways to combine different storage devices. Using an ultracapacitor in conjunction with a battery combines the power performance of the ultracapacitor with the greater energy storage capability of the battery. As with other technologies, the right solution will be dependent on how well it meets the needs of a given situation.

Pumped Hydro

Pumped hydro works by storing gravitational energy in the form of water at a high elevation so that during periods of high electrical demand, the stored water can be released through turbines, generating electricity. The water is pumped from a lower elevation reservoir to a higher elevation during off-peak hours when low-cost electric power is used to run the pumps.

Pumped hydro is currently the most widespread storage method in use at power networks. Generally they use natural bodies of water, but man-made bodies of water can also be created.

Compressed Air Energy Storage

In this storage option, compressed air is stored in airtight underground caverns or former mines. Compressed air energy storage (CAES) utilities can use off-peak electricity to compress air and store it. When the air is released from the underground mine or cavern, the air expands through a combustion turbine to create electricity.

ENERGY DISTRIBUTION

I WAS FIRST INTRODUCED to the importance of energy distribution in an unlikely place. I took a course called Money and You with Robert Kiyosaki at the Excellerated Learning Institute. At the time, Robert was a relative unknown, but now he is the author of many best sellers and teaches seminars to thousands at a time. He was a Buckminster Fuller enthusiast and was—and I assume still is—a proponent of global energy solutions.

I got to know Peter Meisen, executive director of the Global Energy Network Institute, in one of Robert Kiyosaki's Global-Educators programs. I also met my husband, Robert Meredith, there, but that's another story.

COMPONENTS OF ENERGY DISTRIBUTION

Once energy is converted into electricity, it must be transported to where it is needed. There are three main parts to the distribution of energy. The first, power generation, has been covered in chapter 7, on energy supply. The second part is transmission. Transmission facilities do the main job of matching supply with demand. They also ensure the reliability and quality of the energy supplied. They assess how much energy is needed and how to get it where it has to go; they also deal with transmission from the power plants to the substations.

The third component is distribution. This includes the power lines that run underground or overhead to distribute the energy to us whenever and wherever it is needed.

Depending on how each component is accomplished, there can be energy loss in the process. This energy loss needs to be minimized to ensure maximum energy efficiency.

GLOBAL ENERGY GRID

When the lights are on in Australia, most people are sound asleep in Austin, Texas. But the power plants do not go to sleep. Instead, they are producing around the clock. The energy produced in off-peak hours is wasted—unless it is transported to where it is needed. In order to transport energy from Texas to Australia, there needs to be a pathway. There are roadways and airplane routes established if you want to get to or from Australia. Similarly, a grid of interconnected energy distribution lines must be in place to transport energy. Component #6 is an international energy grid.

Electrical grids already deliver electricity throughout developed countries and can similarly be established in developing countries. The remaining technical question is how to connect different continents across large bodies of water. The map shown in figure 9-1 was developed by Buckminster Fuller, noted inventor and futurist.[1] The map is drawn differently than traditional maps; it shows clearly how it is possible to connect the world through the use of a global electric energy grid. The technology exists today to do this.

Traditional maps make it appear that the continents are far away from each other. When seen from this view, it's more apparent how the entire world can be connected. For instance, Siberia is only ninety miles from Alaska. The longest distance between individual islands of the

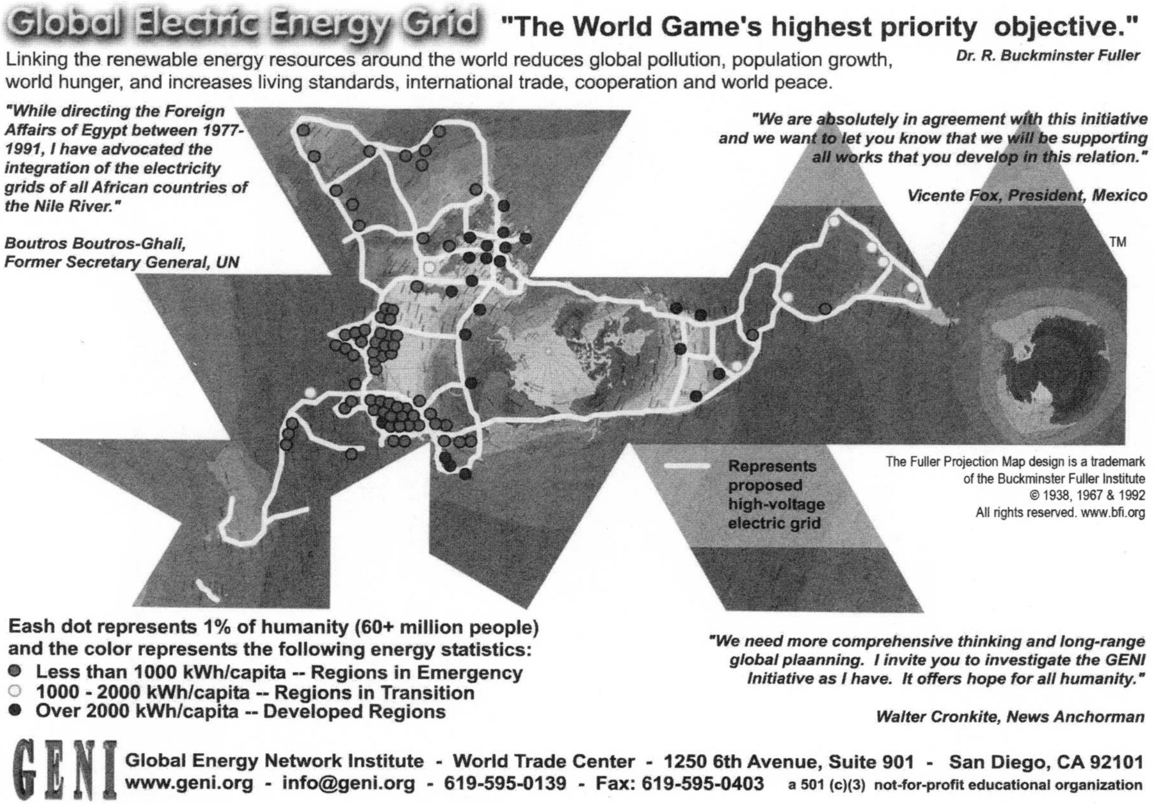
Global Electric Energy Grid
"The World Game's highest priority objective."
Dr. R. Buckminster Fuller
Linking the renewable energy resources around the world reduces global pollution, population growth, world hunger, and increases living standards, international trade, cooperation and world peace.
"While directing the Foreign Affairs of Egypt between 1977-1991, I have advocated the integration of the electricity grids of all African countries of the Nile River."
Boutros Boutros-Ghali, Former Secretary General, UN
"We are absolutely in agreement with this initiative and we want to let you know that we will be supporting all works that you develop in this relation."
Vicente Fox, President, Mexico
TM
Represents proposed high-voltage electric grid
The Fuller Projection Map design is a trademark of the Buckminster Fuller Institute © 1938, 1967 & 1992 All rights reserved. www.bfi.org
Eash dot represents 1% of humanity (60+ million people) and the color represents the following energy statistics:
Less than 1000 kWh/capita -- Regions in Emergency
1000 - 2000 kWh/capita -- Regions in Transition
Over 2000 kWh/capita -- Developed Regions
"We need more comprehensive thinking and long-range global plaanning. I invite you to investigate the GENI Initiative as I have. It offers hope for all humanity."
Walter Cronkite, News Anchorman
GENI
Global Energy Network Institute - World Trade Center - 1250 6th Avenue, Suite 901 - San Diego, CA 92101
www.geni.org - info@geni.org - 619-595-0139 - Fax: 619-595-0403
a 501 (c)(3) not-for-profit educational organization

FIGURE 9-1: Global Electric Energy Grid

Bering Strait is only three miles, so Russia can be connected to Alaska through the Bering Strait.

There are 1.6 billion people in our world who have no electricity, and the map indicates where those people are located. The availability of energy is a primary indicator of material comfort and survivability. When any society is able to provide about 2,000 kWh per capita per year, it can move out of poverty conditions and join the regional/global marketplace.

The technology to create this global grid already exists. Not only would it enable us to be the most efficient with the energy resources on our planet, it would also make it possible for impoverished portions of the world to have a quality of life that has been unavailable to them. Electricity can make this possible.

Energy Security

Just as computer networks need to be protected from computer viruses, electrical grids need to be protected. If one part of the grid were to have problems, the network must have the ability to isolate that section from the rest of the grid.

It is also imperative that redundancies be built into the grid structure. This is like a transportation system: If one route is blocked, there needs to be another pathway to your destination. It is redundant because it achieves the same purpose as the first route. However, the redundancy functions as a backup plan. If you drive a certain route every day and the road is closed one day, having an alternative route is vitally important.

We need to establish redundancies in the global network so that we don't create dependencies that are not beneficial. An international grid enables us to do this. Without energy, virtually none of the other large-scale systems of our society work. Therefore, energy security is an important part of the solution.

OFF-GRID

It may not be economical or even feasible to install transmission lines in some situations. Off-grid applications can handle electricity requirements in remote locations or any other situation where systems need to operate independently.

The shift to off-grid energy is similar to the shift from mainframe computers to personal computers in the '80s. Prior to that, computer systems were made up of a large computer that did all of the processing, and multiple "dumb" terminals were attached to the mainframe computer to be used for entering and viewing the information.

When desktop machines capable of doing their own computing were invented, computer technology began moving toward decentralization. Much of the "thinking" was done at the workstation level. These smaller machines could still communicate with the mainframe computer when necessary, or they could be run completely independently.

Consumers may choose to generate their electricity off-grid even if distribution through the grid is available to them. With new electricity and storage technologies designed for use in residences and small businesses, staying off-grid may be appealing. As personal computers have shown, this is feasible. The benefit to separating from the computer network is that the individual user has total control over the operations. The down side is that there is not a central control area to ensure reliability, availability, and serviceability of the computers; each user has to take care of his or her own.

In the world of energy, staying off-grid has the potential of providing cheaper electricity. However, off-grid means no redundancies in the case of problems. Perhaps the best of both worlds is to stay connected to the grid but to maintain control of local operations. In this way, users can operate independently, but if their system doesn't generate enough power, they can tap into the local power supply so that their service isn't interrupted.

In fact, this is what has evolved for computers. Computers are almost always connected to a network these days. They run applications at the individual computer level, while connecting to the network when it is useful to communicate with mainframe computers on the World Wide Web or in their companies or to communicate with other independent computers through the network.

FEEDING BACK TO THE GRID

As electricity users increasingly use technology to generate and store their own electricity, and elements of the smart grid are implemented, new possibilities will open up. Bidirectional capabilities become feasible. Customers can get electricity from the grid and also provide the grid with electricity.

Once their contribution to the grid can be monitored and tracked, customers may be paid for the electricity they provide. This would compensate for wear and tear on batteries and other equipment. This sort of payment is called a feed-in tariff. It means that when you feed energy into the grid, you get a tariff or fixed payment for the energy that you provide to the grid. Again, it is a method that would allow us to be most efficient and would provide the potential for less-costly energy.

SMART GRIDS

To enhance the efficiency of the system, "smarts" can be built into the grid to feed information back to the source of electricity. To implement this system, addresses would be assigned to locations on the grid so that messages could be sent to and from each location, just as is done with standard mailing addresses used by the post office.

At the moment, the only way the utility companies know when you have a power outage is if you call and tell them. Smart equipment on the user end could collect and send information. This capability could be built into thermostats, computers, appliances, and any other electrical item that accesses the grid. Smarts on the supplier side are needed to collect information and respond to it. If a location is having problems with its electricity, a smart grid could tell the distribution center so that the problem could be located, identified, and quickly fixed. When necessary, parts of the grid could be isolated so that the problem would not spread.

The smart grid could even enable more centralized control for such roles as turning off equipment when not in use. Most computers already have this capability built into them, which allows for computer power management software to monitor and control the equipment from data centers. Energy waste is reduced, thus saving money in electricity costs.

Just as the new technologies for generating electricity will create new types of jobs, these pieces of the grid will create new jobs that need to be filled. There are many opportunities in the global-energy solution.

10

GOVERNMENT INVOLVEMENT

WHEN I WENT TO the Texas legislative session, I was baffled, as I presume virtually any political novice is. I asked questions of anyone who would listen. Fortunately for me, in the gallery of the Texas House of Representatives I met Dwight Harris, a retired schoolteacher who is now a lobbyist. He not only answered my questions, he also walked me over to the library and showed me how to research a bill and how to track it through the legislative process. He told me who the key legislators and staff were that I needed to talk to. He explained what was happening and what to watch for. Though I was a relative stranger, he even loaned me a $20 bill when I didn't have the cash with me to make copies in the library!

This was the last two weeks of the session, which meant the likelihood of getting a bill through the many steps of the legislative process that session was really low unless it was already near completion. To his credit, Dwight didn't discourage me. He is a teacher, and he knows that the best way to learn is by doing.

I visited countless offices of senators and representatives, talking to them about getting a solar-energy addendum added to a bill that was still being debated those last two weeks. Virtually everyone was helpful and also curious. Virtually all of them asked what organization I represented. I replied, "I'm a citizen." They were surprised. I guess not many citizens get so involved. That's too bad.

When I went to the Business and Commerce Committee meeting, a staff member of GalleryWatch, an organization that provides transcripts of legislative sessions, clued me in on when to speak. I must have looked like a doofus as I bumbled my way through the process, but I was passionate and knowledgeable, and as I said at the hearing, "I may not know the legislative process, but I do know energy." After I made my points about the bill, Senator Kirk Watson, a man of great warmth and wit and a former mayor of Austin, told me, "Good point and well said." Did I mention that he's wise, too? Unfortunately, that wasn't enough to make the bill succeed.

Perhaps the biggest thing I learned was how important government policies and standards are in establishing the structures on which our country operates. I also gained great respect for the work these officials do—something I didn't necessarily feel before that experience. And I discovered what a difference my vote could make in electing officials who recognize the importance of focusing on an energy solution.

POLICIES AND STANDARDS

Our government has the authority to initiate solutions to the country's ills. With respect to energy, the government's role can be divided into a few key categories: setting standards and policies, funding research, spurring the development of industries for new sources, education, and international negotiations.

Policies are the rules and courses of action that we agree to abide by. Standards are the accepted performance levels, criteria, and norms by which those actions are judged. For instance, there is a policy of having speed limits for all roadways. The standards are the actual speed limits that are established.

There are policies and standards that are required to support the shift to a balanced energy solution and minimize environmental or social costs. And there are some very smart, market-driven programs already in place.

Cap and Trade

We've established that we want to move to clean forms of energy. One way is through the use of cap-and-trade policies. These are ones in which individual energy producers are allowed to develop their own strategy for complying with a cap, or limit, set for emissions. Producers can meet the requirements through the installation of pollution controls and/or implementation of efficiency measures, among other options. They can also trade allowances, selling them if they exceed the requirements or purchasing them if necessary. The sale or purchase of allowances affects their profits, motivating them to implement controls.

Cap–and–trade policies are currently enacted at the state level, although there have been proposals to introduce them at the national level.

Renewable-Portfolio Standard

The renewable-portfolio standard (RPS) is a way to let the markets drive the move to renewables for electricity generation. The RPS sets the thresholds for the proportion of renewable energy sales that electricity-generating companies are to meet as a proportion of their total sales. They demonstrate their compliance by acquiring renewable-energy credits. For example, if the RPS is set at 5 percent and a generator sells 100,000 kWhs in a given year, the generator would need to have 5,000 credits at the end of that year. Renewable energy credits can be traded in a method similar to that used with cap and trade for emissions.

The RPS is modeled after the federal sulfur dioxide allowance trading program. Because of the high penalty associated with noncompliance for the sulfur dioxide allowance, the EPA has not had to take any enforcement actions. It is far more economical for power plants to comply than not to.

The standard does not restrict investors and generators to a particular type of renewable energy, which technologies to use, price and contract terms, or whom to do business with. They can determine how to meet the requirements, as long as they possess a sufficient number of credits by year's end.[1]

Currently, the RPS is enacted only at the state level, and many states have not done so yet. There are also efforts to enact a federal RPS.

Renewable-Fuels Standard

The Energy Policy Act of 2005 established a renewable-fuels standard as an amendment to the Clean Air Act.[2] It specifies that a certain proportion of fuels must be renewable. This includes motor-vehicle fuel produced from plant or animal products or waste, such as ethanol and biodiesel, as opposed to fossil fuel sources.

Any party that produces petrol gas for use in the United States, including refiners, importers, and blenders, must meet the requirements. There is also a trading program for renewable fuels. The trading program allows obligated parties to comply with the annual renewable-fuel standard through the purchase of renewable identification numbers (RINs) even if they cannot or do not wish to blend renewable fuels into gasoline.

Other Policies and Standards

There are many other policies and standards that both directly and indirectly affect energy solutions. Some examples are: changing the standards that automakers must adhere to on gas mileage; reducing speed limits; establishing policies for developing a hydrogen infrastructure; standardizing units of measure for energy; establishing green-building standards and energy-consumption standards; requiring environmental education in schools; removing or changing subsidies on energy sources to support the movement to clean and renewable energy; and instituting appropriate taxes.

Restrictive policies and standards can block energy flow. Sometimes this is appropriate. Some policies and standards channel solutions toward particular pathways. Sometimes that's appropriate. Perhaps there are some policies and standards that need to be relaxed or removed to allow for flexibility in dealing with the uncertainty of the energy path ahead. For instance, policies that specify requirements for certain types of energy sources may prohibit the inclusion of new discoveries. In all cases, the primary focus must be to move us away from a gloom-and-doom vision and toward a healthier environment for the planet and its inhabitants.

EDUCATION

Many of the pieces of the energy puzzle are in place today, but people may not know about them. There is an abundance of information on the Internet, including government websites.

New developments are continually arising. Find a pipeline for information as new developments occur and as new alternatives become available in your region. Utility companies and other local government entities often provide this pathway, as do local media.

Perhaps the best place to provide education on energy and the environment is in schools. Energy provides a vehicle for students to get involved, and it can motivate them to learn more about math and science—a vital objective for American education.

Education and communication are needed in many different spheres. Keeping good communication flowing enables us to work together.

11

GLOBAL RELATIONSHIPS

MY HUSBAND IS FROM Wales, a beautiful country of hills and valleys. We lived in Australia with its breathtaking beaches—Robert for eighteen years and me for four. I've had the opportunity to visit or work in more than twenty countries, as well as almost all of the states in America. We have a beautiful planet (Robert calls it a garden planet), and we want to keep it that way.

Buckminster Fuller called our planet "Spaceship Earth." He even wrote a book called *Operating Manual for Spaceship Earth* (Lars Müller Publishers, 2008). As Fuller has said, we don't have a crisis of scarcity on our planet, we have a crisis of ignorance. If we understood how the world works, we wouldn't have the problems we have. He believed that the role of humans on this planet is to use our intelligence, and through experiment and intuition, solve the problems that face us. And the solutions must work for everyone.

ELECTRICITY AS A SOLUTION TO WORLD PROBLEMS

Electricity provides the hierarchy of needs for a good life: (1) clean water, (2) food, (3) shelter, (4) sanitation systems, (5) health care, (6) transportation, (7) education, and (8) security. When it is derived from clean energy, it also addresses global warming and the most basic human need: clean air. Yet 1.6 billion people in our world have no electricity! If we want to improve global relations, this is an obvious place to start. It's time we found a way to share the resources of our planet.

Electrical energy can help solve many world problems, including hunger and overpopulation. Electrical energy can supply power for producing and preserving food and medicine and for ensuring clean water. Not only that, but studies also show that when food and health care are readily available, infant-mortality rates decrease.

You might think this would increase overpopulation, but birth rates are shown to decrease as nations develop. Why? Poor families have more children to ensure that there will be someone to care for the parents when they get old. With fewer people dying from hunger-related causes, fewer "insurance births" are required. Providing electrical energy to underdeveloped countries can improve their standard of living and eliminate some of the key problems existing there.

Figure 11-1[1] shows how uneven the distribution of energy is. If 1.6 billion people in our world have no electricity, that's also a huge untapped market. But if those people don't have any money to pay, they can't be part of the global marketplace. For an exchange to occur, each party must have something the other perceives as having value, and they both must have the ability and the desire to carry out the exchange.

This is another reason alternative fuels are an important part of the potential solution. Growing crops requires low-skilled workers. Agriculture in impoverished countries therefore provides a product to trade and a means to start creating their own energy infrastructure. Of course, there are challenges. Research is necessary to grow crops for alternative fuels without wiping out forests. It's also necessary to determine the optimal crops for the terrain. When the commitment is in place to overcome these obstacles, the energy will be invested to make it happen.

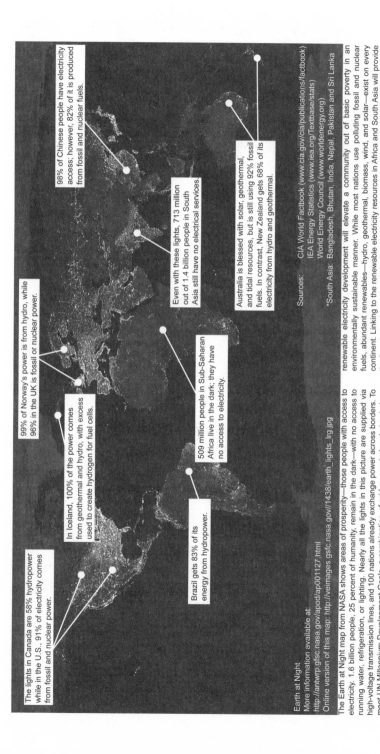

The lights in Canada are 58% hydropower while in the U.S., 91% of electricity comes from fossil and nuclear power.

99% of Norway's power is from hydro, while 96% in the UK is fossil or nuclear power.

98% of Chinese people have electricity access; however, 82% of it is produced from fossil and nuclear fuels.

In Iceland, 100% of the power comes from geothermal and hydro, with excess used to create hydrogen for fuel cells.

Even with these lights, 713 million out of 1.4 billion people in South Asia still have no electrical services.

Australia is blessed with solar, geothermal, and tidal resources, but is still using 92% fossil fuels. In contrast, New Zealand gets 68% of its electricity from hydro and geothermal.

Brazil gets 83% of its energy from hydropower.

509 million people in Sub-Saharan Africa live in the dark; they have no access to electricity.

Earth at Night
More information available at:
http://antwrp.gfsc.nasa.gov/apod/ap001127.html
Online version of this map: http://veimages.gsfc.nasa.gov/1438/earth_lights_lrg.jpg

Sources: CIA World Factbook (www.cia.gov/cia/publications/factbook)
          IEA Energy Statistics (www.iea.org/Textbase/stats)
          World Energy Council (www.worldenergy.org)
*South Asia: Bangladesh, Bhutan, India, Nepal, Pakistan and Sri Lanka

The Earth at Night map from NASA shows areas of prosperity—those people with access to electricity. 1.6 billion people, 25 percent of humanity, remain in the dark—with no access to running water, refrigeration, or lighting. Nearly all the lights in this picture are supplied via high-voltage transmission lines, and 100 nations already exchange power across borders. To meet UN Millennium Development Goals, a combination of grid-connected and stand-alone renewable electricity development will elevate a community out of basic poverty in an environmentally sustainable manner. While most nations use polluting fossil and nuclear fuels, abundant renewables—hydro, geothermal, biomass, wind, and solar—exist on every continent. Linking to the renewable electricity resources in Africa and South Asia will provide the foundation for ending hunger and poverty.

FIGURE 11-1: Electricity Is Essential for Development

There are huge desert areas in some Third World countries that could provide sand for converting into silicon, which can be used to generate solar energy. Assisting these countries in the development of such technologies will help them generate their own energy, which indirectly helps everyone else on the planet.

At about 2,000 kWh per capita per year, a society will move out of basic poverty conditions and is then able to move into the regional/global marketplace. Besides enabling these countries to provide energy to their citizens, this creates a huge opportunity for us to export American goods and services to them.

The most significant point is, we must be a part of this emerging global energy marketplace, or it will happen without us.

SHARING RESOURCES

The relationships between countries are not unlike those you see in businesses, in households, and even on the playground. When two children want the same toy, there is conflict. On a global scale, energy is the toy. With energy, though, there doesn't have to be scarcity. There are more than enough toys to go around.

If we are to ramp up all of these alternative-energy sources, we also want to be efficient with them, to minimize the demand planet-wide. We can't address energy in a vacuum. Finding ways to meet the needs of countries such as China and India is essential if we hope to solve our own needs.

We must retain a sufficient proportion of oxygen-producing land, which means we leave enough land undeveloped. We need to work with Third World countries to ensure that their development, as well as ours, proceeds with this in mind. We don't want to overrun the planet with unrestricted growth.

Limiting growth is necessary on another level, too. We create carbon dioxide through exhalation and methane from human waste. We need to make sure we don't populate this planet to an unsustainable level.

NATIONAL SECURITY

When every kid on the playground has enough toys to play with, there is no need to fight. Working with each other against the bigger threats to the health of our planet can turn our focus away from working against each other. Therefore, working together on a global energy solution can boost national security.

Note in the list at the start of this chapter that education has priority over security. Education reduces fear. Education overcomes helplessness and frustration. Education empowers others to contribute to solutions.

A crucial element, then, is to educate other countries and to be educated by them and to establish the trust to make those relationships work. Education can also help all countries understand the impact of their actions on each other and determine how to progress in the most effective ways possible for the entire planet. No country can consider itself separate from the rest of the world.

12

THE TRANSITION

CHANGE CAN BE MESSY. My family moved to a new home recently. Our old home was nice, and we like to think we're even better off now, but it got chaotic in between. We didn't worry about the fact that it was chaotic, though. We were committed to the transition, so we stayed focused on that goal. We even tried to have fun with it sometimes. That's the way I think about this energy transition. We can't just stop in the middle and say, "Oh my, this is a mess!" because we have to go through that stage to get to the better future. We may face challenges figuring out where everything belongs, but we'll get there. Change is inevitable, but struggle and suffering aren't.

SYSTEMIC CHANGE

Now that we've covered all of the components of an energy solution, look once more at the headings in this book. If we're to be successful resolving global warming and the energy crisis, each of the issues addressed in this book must be resolved. That requires systemic change, meaning all of the systems must change in a complementary way.

An example of systemic change is making the units of measure we use in the United States conform to those used in the rest of the world—meters and centimeters instead of feet and inches, grams and kilograms instead of pounds and ounces. To make the change, though, we could not simply start using meters and grams. The makers of rulers and measuring cups couldn't simply change their markings, with the rest of us then going out and buying one of each. We would need to change our textbooks and our cookbooks. We would need to change architectural drawings and engineering drawings for every product that is manufactured. Over time, we would need to change the design of the parts themselves so that they could be measured in even multiples of meters and grams. And we would need to ensure that we time the changes of one set of parts to match the changes in parts that they connect with.

Again, our five objectives are to:

- reduce dependence on foreign oil
- minimize global warming
- stimulate the economy
- promote global relationships
- reduce world hunger

The first requirement is to commit to making the changes necessary to achieve these results. As said in the first chapter, without appropriate energy, our commercial, educational, health, finance, transportation, and defense systems cannot operate. These systems will change. We need to consciously choose to change them in a direction that works.

In commercial systems, particularly in the new fields related to energy, the environment of competition must give way to one of cooperation, in finding ways to share knowledge and expertise. The primary goal for alternative vehicles, fuels, and sources of electricity must change from maximum profits of the individual companies to maximum energy produced by the collective group. The primary goal of the nonrenewable companies must shift to reduction in toxicities produced per unit of energy type provided. In both cases, a measure of success is how quickly and efficiently they are able to make these changes.

Educational systems must teach about the environment. In a December 2007 article in the *Austin American-Statesman*, Brian Day, executive director of the North American Association for Environmental Education, said, "Young people are graduating from high school totally environmentally illiterate" ("Schools fail to teach children about their world"). Adults also need an avenue for better environmental education.

The primary goal for health systems is not to keep people alive, but rather to return them to the ability to give and receive energy so they are a contributing member of society. This includes personal energy management on all levels—physically, mentally, and emotionally. Rather than allowing patients to believe that a health care provider is primarily responsible for fixing them, the systems must instill personal responsibility in each person for their health. Our personal energy is key in global energy solutions.

Financial systems must more adequately recognize the value of intellectual capital and human energy as well as money and other forms of capital, particularly in the area of new energy solutions. There are many solutions that have received years of unpaid personal investment. There are those who have been storing energy in the form of money. These two factions need to merge to create the solutions that are so crucial now.

The whole idea of defense needs to be revisited. Defense implies an ongoing state of conflict. Conflict implies opposing goals. Sometimes these conflicts are necessary to set a third pathway that is better than either of the two. But we are on this spaceship together. Bombing one side is not conducive to the smooth flying of a ship. We need to instead focus on International Conflict Resolution and how to provide a distribution of energy that works for all.

The media must recognize their responsibility in driving the moods of our society around energy. Whenever possible, when presenting the problems, they need to follow with solutions so the momentum is left in a positive direction. Report on successes. Rather than pundits looking for why things won't work, put a focus on what it would take for success.

In addition, let's look at the energy-related systems we've been specifically addressing.

Transportation

The use of all-electric and the various hybrid plug-in cars with alternative fuels would provide a bridge from the current primary sources of energy for transportation without requiring a significant change from energy distribution systems, since gas stations and electricity grids already exist. But to fully resolve our current energy and environmental problems, there are other challenges that must be faced. Remember the lemonade stand? If you sell one glass or one hundred, you still have to spend money on certain supplies, including the pitchers, the sign, the table, and the chairs. The first few glasses seem really expensive. Once you start making lots of sales, though, those up-front costs seem worth it. The same is true for new vehicle technologies. In addition, these new technologies require early funding for research and development to fine-tune the design of vehicles and to make them financially viable.

Besides covering the initial costs, manufacturing systems must also become increasingly efficient to ensure profitability. As they say, practice makes perfect. As businesses practice making their products, they get more efficient, which drives down the cost to produce the products, which drives prices down. The quality of their products improves. But new businesses don't get to practice unless there is enough demand. But demand won't go up until the prices become better than, or at least comparable to, the other options. In addition, the systems to service the new technologies need to be in place. Until there are enough customers, it isn't cost-effective to set up service centers.

So if those are the problems, what's the solution? One strategy used by alternative-vehicle companies is to sell fleets of vehicles. This helps spread the costs over more vehicles, plus it justifies and pinpoints where to set up service stations. If groups of people band together to buy products, they can also help create the market and create a successful transition plan for all involved.

All of the above considerations apply to hydrogen vehicles, *plus* the hydrogen station infrastructure must be installed. Whether the hydrogen initial public offering auction is the method, the support structure must be phased in over time.

A phase-in is required for alternative fuels as well. The introduction of alternative fuels at stations needs to be timed to match the phase-in of compatible vehicles.

The common theme through all of these is that there must be a commitment to make the changes and there must be coordination of those changes. Policies and standards can help instill the commitments. Investors play a key role in financing these transitions. There are risks, but there are also great potential rewards.

Plug-in cars, hydrogen cross-country buses and trucks, and supporting infrastructure can provide the foundation for a stable transportation-transition strategy so we can still go on cross-country road-trip vacations! This planet is meant to be enjoyed, and we must keep that in mind as we make the changes.

Electricity

The transition for systems that supply electricity and other nonmigratory energy is affected by the same sorts of considerations as transportation fuel. Prices rise when demand exceeds supply, as is already happening with petrol gas. When there is a finite supply, as is also true for natural gas, coal, and nuclear power, the question isn't if supply will run out; the question is when. The increase in energy demand is occurring, both in the United States and other countries, so eventually the price of scarce resources goes up, which makes the alternatives—in this case the renewable-energy sources—more feasible and more critical.

Of course we don't want our energy prices to go up, because it will affect the economy, and remember, a good economy is one of the criteria for a good solution. The question becomes how to make the transition as painless as possible. If the ramping up of renewables is timed such that the prices go down before the supply of oil and gas dries up, our standard of living doesn't have to be negatively affected by the cost of energy.

That's a big if, but it's possible. The technology already exists for all of our electrical-energy demand to be supplied from clean, renewable sources. The challenge is in making them economically viable and in ramping up the demand quickly.

Supporting Systems

As the use of unpredictable (wind and solar) electricity sources increases, storage will also become more critical. Improvements in energy distribution are also vital for matching supply with demand in the most efficient way.

Policies, standards, and controls will need to continue to evolve so we don't create more problems by raising emissions through increased energy consumption and we ensure that our children don't get into the same trouble from heavy dependence on a nonrenewable energy source. The use of green-building and green-living principles is also crucial for ensuring a quick and effective transition.

The specifics of the transition remain to be seen. While we transition, we need to be aware that the way we measure our success needs to adjust, too. We need to look at the metrics differently.

REGIONAL CHANGE

For the most part, changes can and probably should happen regionally. Each region can prioritize its efforts based on the natural resources available. To be successful, each region must address all of the components of an energy solution: plug-in vehicles, alternative fuels, hydrogen for cross-country vehicles, clean and renewable energy sources, storage, and a grid.

We might be surprised how, with the right focus, we can very quickly change. Austin Energy had a plan for 20 percent renewable energy by 2020. It reached 6 percent by 2007 and has commitments from suppliers that will allow it to reach 10 percent by the end of 2008. On February 7, 2007, Mayor Will Wynn announced a new goal of 30 percent renewable energy by 2020.

Of course, there will be a learning curve involved. We wouldn't expect a high school basketball team to play as well as the pros. Neither should we expect new technologies to work as well as those that have been refined over many years. What we should expect is for renewable-energy suppliers to correct problems as quickly as possible. The ones that do will move on to the pros.

Each region can also focus on green building, green living, and carbon offsets for their area and consider how to assist others in their region with the transition. If people in your region are struggling, the region as a whole is affected. If people in your region aren't doing their part, the region as a whole is affected. When the region is healthy and has plenty of energy of all types for itself, it will then have the ability to supply it to others.

Each region must take care of itself, while maintaining an eye to how it fits into the bigger picture, and how to exchange energy with other regions for the benefit of both.

PERSONAL CHANGE

Hopefully, this book highlights that it isn't a matter of which part of the energy picture needs to change; all parts need to change, including us. What if we see brownouts and blackouts during the transition? If it happens, the first question to ask ourselves is how we can reduce or even out the demand through our practices and purchases and by using storage to create a buffer. As we move to plug-in vehicles, we need to develop the habit of charging our vehicles at night so we don't exacerbate peak demand.

We all need to have patience and understanding along the way and continue to look for ways to personally contribute to the solution. Observe where your own energy is focused. Is it on "woe is me, how awful this is" or is it on "what will it take to make things work"? Get in motion to find answers rather than sitting and waiting for someone to bring them to you. When you do get in motion and you encounter obstacles, look for solutions rather than who to blame. And we must also recognize that it isn't our time that has the value; it's our energy. We can't wait for someone else to fix the situation. We are each responsible.

If you are personally challenged during the transition, the best thing to do is to find a way to be part of the solution. Resisting wastes energy.

Choose a direction. Sitting on the fence is the most uncomfortable place to be.

Opportunities

If you want to get involved directly, such as getting a job in a "clean" industry, pick the part of the puzzle that is most interesting to you. If you can't decide what to focus on, check what you feel most connected to—taking into account the people, the mission, and the work itself. Study math and science. They are the languages that describe the physical world. See how you can best use your skills and knowledge. Then choose what interests you. There's plenty to choose from.

With the new technologies, new products, and new services, there are so many opportunities for new careers! Job growth will take place in the development of the new energy sources, in vehicle technologies, and in energy storage and distribution. The technical complexity of the jobs will cover a broad spectrum, from research and design at one end to installation and manufacturing at the other.

Some of these new careers will represent a natural evolution from workers' previous career paths. For instance, working with solar products requires skills and knowledge similar to those used with computer products. Hydrogen technology involves a distillation process similar to oil distillation.

Instead of waiting until jobs become plentiful in the new fields, take the initiative to learn about them in any way and to whatever level you can. Learning is never wasted.

Our Human Energy

Probably the most important thing we can all do is to commit to investing our personal energy in creating the solution.

In the physical world, the term *energy* defines the capacity of a system to do work. Einstein's most famous equation, $E = mc^2$, established the equivalence of energy and mass, or matter. Don't get put off by the formula. You can understand it. Really. You live it every day.

Math is about relationships. This equation says that energy is mass multiplied by the speed of light times the speed of light. The way the equation is written leads us to think that it means we could get a large amount of energy from a small amount of mass, if only we could access it. However, a lesser-known fact is that the original equation was written as $m = E/c^2$. This implies another focus to the mathematical relationship: that a small amount of mass can be created from a large amount of energy.

What does that mean? It means we convert energy into mass in the physical world by concentrating energy on its creation. We create a company and business results from the marshaling of energy. We create any results through our deployment of energy.

Thoughts have energy. Words have more. Actions have even more. Results have more again. If you keep putting your thoughts, words, and actions toward a goal, you create results. Mass. Physical, tangible results.

Now let's look at an example of how to apply this to energy management. Let's apply $m = E/c^2$ to examples of using our human energy to address global energy.

Thoughts lead to words lead to actions lead to results. We can each start with the thought "I wonder how I can help with this global energy management thing" or "I wonder how much money I can save by reducing my energy use" or "I wonder how much energy our organization can save." Then add personal energy to that thought. Talk about it with friends. Turn it into a goal. Make it a game, even. To make it more real, write it down. Now you've used energy to create an initial form of mass for the idea.

You might, for example, write, "I will reduce my energy usage and my carbon footprint by X kilowatts per month." Then take action to advance your goal. Notice the results—physical, tangible results. See how many kilowatts you use on your energy bill. Set a goal of at least a 20 percent reduction from your current bill. See how the kilowatts change. You'll be surprised at the results. The tangible figure will confirm that you have made a difference. And it will save you money!

13

WHAT WE ALL CAN DO

ON THE MORNING THAT I was to speak at a renewable-energy conference in Fredericksburg, Texas, we learned that one of my husband's very best friends had died. Jeff was one of those people who had no pretense—what you saw was what you got. At 15, he was diagnosed with cancer. He beat it, but it cost him an arm. Yet shortly after meeting him, you would forget that the arm was missing. He didn't let it stop him from doing anything. He invented a knife/fork so he could cut his own food with one hand. He invented a bow he could shoot with one arm. He was a ski instructor-examiner, and experts marveled that, although he was missing one arm, there was no difference between his turns to the left or the right. He was an adventurer. He was a loving husband and father.

We had been planning for Jeff and his wife, Becky, and daughter, Libby, to attend a Robert Earl Keen concert in Austin. We never got around to it. In June Jeff finished a round of chemo and radiation. It looked like he had beaten a new cancer that had appeared in his body. But in September, he was gone. We thought we had all the time in the world. We were wrong.

As I prepared for my talk, I thought of how Jeff's death related to my life. Years before, I had started researching how laws of nature apply to human nature. I had written about the subject but hadn't done much with it. I realized time was passing for me. Time is passing for all of us. Jeff's death was a wake-up call for me.

As I was writing this book, I kept thinking, "Who am I to write a book about energy?" But I also knew that at the very least, it would educate me and, in doing so, remove my vague uneasiness about the situation. It gave me a place to focus my energy, and I could see that it might even provide value. It was something I could do, so why not try? I'm glad I did.

Each of us has a part to play in the solution. We have to apply our knowledge, interests, connections, and abilities. The lists in this section are intended to be a starting point. I hesitate to make the lists because I may forget important considerations, and others will come into existence as soon as the book is printed. I know some of these things are already happening. I also know that we all can do more, and we each need to find our role in the solutions.

GOVERNMENT'S ROLE

Setting Policies and Standards

- Automobile-efficiency legislation
- Automobile-emissions legislation
- Automobile safety standards
- Standards unique to plug-in vehicles
- Additional hydrogen-vehicle standards, including for the distribution infrastructure
- Reduced interstate speed limits
- Flex-fuel-compatibility legislation for new cars
- Renewable-portfolio standard in all fifty states

- Cap and trade for emissions
- Motor-fuel standards

Funding

- Renewables research and implementation
- Global electricity distribution grid
- Hydrogen infrastructure
- Education

Creating the Industries

- Replace federal government vehicles with electric, plug-in hybrid, and hydrogen vehicles
- Install electric plugs and hydrogen fueling stations at dedicated government gas stations
- Team with post office and state and local government for more electric and plug-in hybrid vehicles
- Use renewables in government buildings
- Institute hydrogen initial public offering auction

Education/Communication

- Develop an energy plan
- Publicize it domestically and internationally
- Promote environmental education in schools

Global Relationships

- Sharing knowledge
- Sharing our plans
- Creating trade agreements
- Developing global distribution grid

COMPANIES' ROLE

All Companies

- Increase energy-efficiency profile of products
- Reduce the emissions profile of products
- Incorporate energy-efficient products and practices in operations
- Reduce/reuse/recycle
- Computer-power management
- Energy storage
- Educate employees
- Transition to hydrogen vehicles for cross-country distribution
- Transition to electric and plug-in hybrids for local fleets

Fossil Fuel Companies

- Reduce toxicity of products
- Invest in hydrogen-technology research
- Develop hydrogen-station infrastructure

Car Companies

- Increase mileage
- Reduce emissions
- Incorporate flex-fuel capabilities
- Produce electric cars, plug-in hybrids, and hydrogen vehicles

Clean- and Renewable-Energy Companies

- Ramp up as quickly as possible
- Improve affordability
- Improve design and quality
- Make it easier for utilities and consumers to purchase renewable products

Construction Companies

- Increase awareness of green-building techniques
- Increase green-building skills
- Use energy-efficient products
- Use low-emissions products
- Educate customers
- Make it easier for consumers to purchase clean- and renewable-energy products

EVERYONE'S ROLE

- Reduce/reuse/recycle
- Join green-energy programs
- Walk, bike, use public transportation
- Share rides
- Incorporate green-building solutions into existing and new homes
- Mow lawns less often
- Xeriscape
- Drive the speed limit
- Turn off vehicle when waiting in the car for someone
- Wear clothes more than once before washing
- Hang clothes to dry
- Combine errands to decrease driving
- Avoid unnecessary travel
- Buy energy-efficient products
- Buy low-emissions products
- Buy alternative transportation: plug-ins and flex-fuel cars, alternative fuels
- Buy carbon offsets
- Invest in solutions
- Plant trees

- Invest in clean- and renewable-energy businesses
- Talk about energy efficiency and solutions to others
- Compliment people on their efforts
- Vote for politicians who support solutions
- Keep learning

14

CONCLUSION

ON THAT SAME TRIP to Fredericksburg, my husband and young son and I camped at Enchanted Rock State Park. In the hectic pace of daily life, I too often forget how lucky we are to live on this garden planet. Reconnecting to Mother Nature reminds me.

Making these changes doesn't have to be gloom and doom and an overwhelming burden. It doesn't have to be struggle and sacrifice. In fact, enjoying the journey is one of the keys to success, because our emotions have energy, too. This planet is meant to be enjoyed.

Sometimes the simple things provide the most enjoyment: taking a walk, reconnecting with friends, growing a garden. Sometimes the greatest enjoyment comes from challenging ourselves to do something we haven't done before.

To find answers, I recommend asking Mother Nature. Mother Nature speaks to us, when we listen. Sometimes the answers come in amusing ways. I have had answers come from the positioning of trees, opening a book at the right time, talking to a complete stranger, watching animals, exercising, and journaling. Of course, it's up to me to interpret what I experience. Each of us must find our own answers relative to the time and place.

If we each ask what we can do for the highest good of all concerned, and ask how to make it personally enjoyable and fulfilling in the process, we can find answers, find balance. We can enjoy the journey and the destination.

WE CAN DO IT

I started by asking what would be possible if we could

- reduce dependence on foreign oil?
- minimize global warming?
- stimulate the economy?
- promote global relationships?
- reduce world hunger?

The solution presented here includes six key components: plug-in vehicles, alternative fuels, hydrogen cross-country trucks and buses, clean and renewable energy sources, storage, and a global energy grid. I have highlighted the need for policies and standards to support the transition. I have addressed the potential of longer-term solutions using hydrogen and nanotechnology. And I have emphasized the role each of us plays in reducing energy demand and emissions through green building, green living, and our purchasing choices.

Chapter 2 (Dependence on Oil) showed how plug-in hybrid vehicles, alternative fuels, and large-scale hydrogen vehicles can eliminate our dependence on foreign oil.

Chapter 3 (Global Warming) addressed the four key areas affecting global warming, and chapters 4 (Energy Demand), 5 (Green Building), 7 (Energy Supply), 8 (Energy Storage), and 9 (Energy Distribution) showed how we can improve in those four areas.

Chapter 6 (The Economy) explored how we measure our economy and the impact that it has on the economy on a larger scale. Chapters 4 (Energy Demand), 5 (Green Building), and 13 (What We All Can Do) and the rest of the book provided a wide range of activities to improve our personal economic situations, through ways of saving energy and money and through our purchasing choices. For those who want to get actively involved in creating the solution, and those who will have the necessity of finding new ways to generate income as we change the ways we generate, store, distribute, and use energy, virtually every chapter of the book provides opportunities.

In chapters 11 (Global Relationships) and 9 (Energy Distribution), global relationships and world hunger are addressed, showing how a balanced distribution of energy can solve world problems to a level never before achieved.

Chapter 10 (Government Involvement) addressed the important role our government plays in creating structures that create rather than restrict movement toward an energy solution.

By putting our personal energy into creating a positive result, we may be surprised by how quickly and effectively we are able to implement a successful solution.

Buckminster Fuller's World Game posed the question: How can we make the world work for 100 percent of humanity in the shortest possible time through spontaneous cooperation without ecological damage or the disadvantage of anyone?

This is the goal. We can do it. Not just survivability but thriveability for all.

Tell me what you're doing and what difference it has made. I would love to hear. Send the results to energyresults@humanexcel.com.

APPENDIX
ASSUMPTIONS USED IN DATA ANALYSIS

Milap Majmundar, TEMBA class of 2006, McCombs School of Business, University of Texas–Austin

Average miles driven per year	12,200	Based on 2003 U.S. census data for all automobiles: cars, buses, pickups, trucks (table 1084)
Percent reduction in average miles driven per year	2.0 percent	As the number of autos keeps rising beyond the number of registered drivers, each auto is individually driven less even though the average miles driven per individual remains the same
Average miles per gasoline gallon for gas-only vehicles	17	Based on 2003 U.S. census data for all automobiles: cars, buses, pickups, trucks (tables 1084 and 1085)
Percent improvement in average miles per gallon	1.5 percent	Due to improvements in vehicle efficiencies and to the increased use of smaller vehicles that burn less fuel

Average miles per gallon for gas/electric hybrid (not plug-in)	55	Estimated mileage for 2006 Toyota Prius
Percent improvement in average miles per gallon	0.5 percent	Electrical per electronic components already very efficient, so less scope for efficiency improvement
Reduction in fuel efficiency of E85 over gasoline	25 percent	Estimate based on data from www.fueleconomy.gov
Average miles per gallon for ethanol/gas hybrid	12.8	
Percent improvement in average miles per gallon	0.5 percent	Hard to judge, but we assume some improvement
Average kWh per mile for electric plug-in	0.3	Average for four vehicle types: compact = 0.25, midsize = 0.3, small SUV = 0.38, large SUV = 0.46 (estimates from Austin Energy)
Avg miles per kWh for electric plug-in	2.88	
Percent new vehicles registered in prior year total vehicles	7 percent	Estimate based on U.S. Auto Industry Report, U.S. Dept. of Commerce, June 2005

Percent attrition of vehicles	3.0 percent	Based on an overall vehicle growth rate equal to 0.75 of U.S. GDP growth rate (the logic is that overall vehicle growth rate would be somewhat less than GDP growth)
U.S. GDP growth rate	5.3 percent	Based on average of ten years of GDP growth from 1998 to 2007 (source: BEA, U.S. Dept. of Commerce)
U.S. inflation rate	2.50 percent	*Wall Street Journal*
Cost of gasoline per gallon in 2005 (pre-Katrina)	$2.10	
Annual increase in gasoline price per gallon (short term)	22 percent	Reflects rapid gasoline price increases from 2005 to 2008
Annual increase in gasoline price per gallon (long term)	10 percent	
Growth rate deceleration from short term to long term	80 percent	
Price of E85 fuel per gallon in 2005	$1.95	Estimate from www.cleanairchoice.org price postings
Annual increase in ethanol price	10 percent	Same as long-term increase in gasoline price
Price of electric power per kWh in 2005	$0.10	Austin Energy

Probability Density Function of commute distance

Probability	Distance (miles)
9.00 percent	0 to 5
19.50 percent	5 to 10
32.00 percent	10 to 15
17.00 percent	15 to 20
7.50 percent	20 to 25
9.00 percent	25 to 30
1.00 percent	30 to 35
0.00 percent	35 to 40
0.00 percent	40 to 50
1.00 percent	50 to 65
4.00 percent	> 65
100.00 percent	

Median	24
Mean	32
Prob < 30	94.00 percent
Prob > 30	6.00 percent
Mean > 30	65.83
Ratio	2.06

Source: Institution of Transportation Studies, University of California–Davis, 2005

Estimate from Austin Energy

Median	25 (fairly close to above data)
Mean	34

NOTES

Introduction

1   Thomas Eisenmann and R. Matthew Willis, "Fuel Cells: The Hydrogen Revolution?" Harvard Business School Case 9-804-144, March 30, 2004; and the "National Hydrogen Energy Roadmap," published by the U.S. Department of Energy, November 2002.

Chapter 2

1   www.eia.doe.gov/oil_gas/petroleum/data_publications/petroleum_supply_annual/psa_volume1/psa_volume1.html.

2   Matthew R. Simmons, *Twilight in the Desert,* John Wiley & Sons, 2005.

3   www.eia.doe.gov/oil_gas/petroleum/data_publications/petroleum_supply_annual/psa_volume1/psa_volume1.html.

4   See Appendix for assumptions for the charts in figures 2-3 through 2-8.

5   For more information about gas/electric plug-in hybrids, visit www.pluginpartners.org.

6   See Appendix for assumptions used.

7   For more information about gas emissions, visit www.epa.gov/otaq/renewablefuels.

8   For more information on flex-fuel vehicles, visit www.fueleconomy.gov/feg/flextech.shtml.

9   Based on average increase of gasoline prices of 10 percent in the long term.

10   For more information about fuel cell technology, visit www.fuelcellpartnership.org/fuel-vehl_buses.html.

11   Our capstone team included Brian Giles, Mike Herrold (who came up with the idea), Sam Heywood, Alex Newton, Julie Simon, and myself.

12   The winning capstone team included Daniel Bounds, Scott Francis, Ryan Franks, Milap Majmundar, Mark Oberhauser, and Neil Peterson.

Chapter 3

1   Energy Information Administration, "Emissions of Greenhouse Gases," www.eia.doe.gov/environmental.html.

2   www.carbonfootprint.com.

Chapter 4

1   For more information about Energy Star, visit www.energystar.gov.

Chapter 5

1   Information on all of these ratings is available at www.eere.energy. gov/consumers.

Chapter 7

1   www.eia.doe.gov/fuelelectric.html.

2   www.biogasworks.com/http://www.biogasworks.com/.

Chapter 9

1   The Buckminster Fuller Institute holds the copyright to this map. It is available through the Global Energy Network Institute at www.geni. org/globalenergy/shop/index.shtml.

Chapter 10

1   For more information about RPS, visit www.awea.org/policy/rpsbrief. html.

2   For more information about the renewable-fuels standard appended to the Clean Air Act, visit www.epa.gov/otaq/renewablefuels/.

Chapter 11

1   Map from apod.nasa.gov/apod/ap001127.html with text from www. GENI.org.

THANKS

The Texas Evening MBA (TEMBA) class of 2006 at the McCombs School of Business of the University of Texas–Austin provided some of the inspiration for this book. Many thanks to them for the hundreds of hours of research they did.

Special thanks to Milap Majmundar, who continued to contribute to this research well beyond the completion of the TEMBA project. His assistance was invaluable.

Thanks to Britt Freund, director of the TEMBA program, who set up the competition, as well as to the other judges: Rob Adams, Director of Moot Corp, Austin Venture Partners; Bob Barnes, president of the International Bank of Commerce; and Betty Otter-Nickerson, chief operating officer of the Lance Armstrong Foundation.

Many thanks to others who shared their energy, including: Mark Kapner, Strategic Planning Office of Austin Energy; Peter Meisen, Global Energy Network International; and many professors from the McCombs School of Business.

Thanks to the staff at Greenleaf Book Group for committing their energy to improving what I had created.

Thanks to my parents, Bill and Mary Kenney, and my six siblings, and so many others who have contributed to my life and this book, especially Melanie Kelly, and Terry and Donna Lipman, who encouraged me when I needed it most.

Lastly, thanks to my husband, Robert, and my son, Ian, who put up with my nighttime inspirations and my sometimes distracted and grumbly creative process. They have been my most important inspiration and support.

INDEX